AF412176

The Competitive Advantage of Regions and Nations

To Rebekka and Aira

The Competitive Advantage of Regions and Nations

Technology Transfer Through Foreign Direct Investment

BORIS RICKEN AND
GEORGE MALCOTSIS

GOWER

Gower Applied Business Research
Our programme provides leaders, practitioners, scholars and researchers with thought provoking, cutting edge books that combine conceptual insights, interdisciplinary rigour and practical relevance in key areas of business and management.

Published by
Gower Publishing Limited
Wey Court East
Union Road
Farnham
Surrey, GU9 7PT
England

Gower Publishing Company
Suite 420
101 Cherry Street
Burlington,
VT 05401-4405
USA

www.gowerpublishing.com

British Library Cataloguing in Publication Data
Ricken, Boris.
The competitive advantage of regions and nations: technology transfer through foreign direct investment.
1. Technology transfer – Economic aspects. 2. Investments, Foreign.
I. Title II. Malcotsis, George.
338.9'26–dc22

ISBN 9781409402381 (hbk)
ISBN 9781409402398 (ebk)

Library of Congress Control Number: 2010937744

Printed and bound in Great Britain by the
MPG Books Group, UK

Contents

List of Figures

List of Tables

List of Boxes

About the Authors

Boris Ricken went to school in Germany and the United States. He studied Political Science together with Economics and Law in Germany and Spain and graduated with a Masters in Political Science from the University of Constance, Germany. Subsequently, Boris earned a Doctorate in Business Administration from the University of Zürich, Switzerland where he also worked as a research assistant. He has performed research in the fields of organization studies, strategic management, social network analysis, knowledge and technology management and foreign direct investment. He has published various books and several peer-reviewed papers for international conferences. In addition, he is an external lecturer at Edinburgh Napier University in the UK, where he teaches a Technology Transfer and FDI module on the Investment Promotion and Economic Development M.Sc. programme. Dr Ricken has gained practical experience related to foreign direct investment and technology transfer whilst employed at a Swiss investment promotion agency. Following this he worked for four years as a Senior Strategy Consultant at the world's second-largest building materials supplier. The company employs some 80,000 people and is present in more than 70 countries. In this position, Boris Ricken was responsible for foreign direct investment projects and business planning in Latin America and Asia. Boris Ricken is now in charge of internal projects and process management at PPCmetrics AG, a leading Swiss consulting firm for institutional investors.

Dr Boris Ricken
PPCmetrics AG
boris.ricken@ppcmetrics.ch
www.ppcmetrics.ch

George Malcotsis was born in Greece and grew up in Egypt. He graduated with a Bachelors and Masters in Engineering from the Universities of Surrey and London. He earned a Doctorate in Engineering from the University of Cambridge, England. He has attended a Senior Executive Business Administration Program at the Massachusetts Institute of Technology (MIT), Boston, USA. George began his career in a teaching and research position at the University of Cambridge. From 1976 to 1990 he worked for a major internationally active engineering consulting firm in Switzerland. His last position in the firm was that of president. In 1990 Professor Malcotsis joined KPMG Switzerland as a partner at director level. His first responsibility was to head the international management consulting division. From 1997 to 2007 he was assigned as managing director of SOFI, the Swiss Organisation for Facilitating Investments, which was an instrument funded by the Swiss Government and operated by KPMG. In 2007 he was nominated visiting professor at the Napier University Business School in Edinburgh, Scotland. His special field of research and teaching is business planning and intercultural business negotiations. Professor Malcotsis' fields of expertise are corporate strategy and business planning, project evaluation, management training and coaching, international corporate finance and mergers and acquisitions. Through managing business operations in over 80 countries, he has gained extensive international experience and an ability to operate across cultural barriers. In 2007 he became Partner of DIAS.

Professor Dr George Malcotsis
International Institute for Investment Promotion, DIAS Management
george.malcotsis@dias-management.ch
www.iiip.ch, www.dias-management.ch

Preface

There is little doubt that technology is of crucial importance for the competitiveness of companies, regions and entire economies. Technology is a determining element for firms and nations to increase productivity, to compete and to prosper. Today we see many developed and prosperous countries that were quite poor only about some decades ago:

- In Asia, Singapore and Korea have managed to increase their gross domestic product (GDP) by approximately six times since 1980.

- The Chinese economy has grown at an annual rate of approximately 10 per cent over the same period.

- In Latin America, Chile's economy has achieved an average annual increase of 4.3 per cent per year over the last 30 years.

- In Eastern Europe, Hungary has increased its GDP per capita from US$3,190 in 1990 to US$12,400 in 2009. Over the same period, the Polish GDP per capita has grown from a mere US$1,600 to US$11,100.

This impressive economic progress cannot be explained by increases in capital or a growth in the supply of labour alone. Rather, it is to a considerable degree driven by the acquisition and usage of new technologies such as machinery, management practices, or production techniques, all contributing to the overall productivity of an economy.

However, there is still a large gap in the technological advancement of nations. On the one hand, industrialized economies such as the United States, Japan, Germany or Switzerland are at the forefront of developing new technologies and applying these. Large shares of GDP in these countries are dedicated to research and development (R&D) of new products or process technologies. On the other hand, many emerging economies in Asia, Africa, Latin America and Eastern Europe have not yet developed their own R&D capacity. They rely on the transfer of existing technology from other countries in order to increase their productivity and competiveness. One major channel of this transfer is foreign direct investment (FDI). Empirical studies have found positive and significant technology spillovers stemming from inward FDI. One of the major reasons is that FDI usually takes place by multinational corporations (MNCs), which exhibit a high ratio of R&D.

The importance of technology transfer via FDI implies that countries cannot simply sit and wait until new technologies arrive in their domain. Rather, companies, regions and nations need to systematically manage the identification, assessment, attraction, absorption and application of new technologies. In this book we want to provide a step-by-step guide on how to manage this entire process of technology transfer in foreign direct investments. The target audience of this book therefore consists of the following groups:

- Managers employed at *investment promotion agencies*. Almost every country has at least one such agency and several have subnational local development agencies. The World Association of Investment Promotion Agencies (WAIPA) alone brings together over 250 investment promotion agencies. This book wants to provide managers of these agencies with the relevant concepts and methods to identify, assess, attract and absorb foreign technologies through FDI.

- Managers of *government bodies* such as investment boards, ministries of economics or ministries of science and technology, active in the field of foreign direct investment and technology transfer. This book wants to support them in setting the appropriate policies and measures suitable to facilitate the economic and technological development of their country.

- Managers working in *companies* active in the field of foreign direct investment, such as local competitors, suppliers, clients, joint venture partners, or wholly owned subsidies of multinational enterprises. Especially in emerging markets, firms strive to absorb new technologies, knowledge and practices in order to gain a competitive advantage. This book aims to provide these managers with a deeper understanding on the determents of a MNC's decision on an FDI project as well as on the processes and mechanisms of technological spillovers to local companies.

- Managers being employed at *international organizations*, development banks and technology transfer intermediaries, for example the World Bank, Asian Development Bank, European Bank for Reconstruction and Development (EBRD), United Nations Conference on Trade and Development (UNCTAD), United Nations Industrial Development Organization (UNIDO) etc. targeting issues of economic development, FDI and technology transfer

- *Students* in programmes related to economic development, international economics, investment promotion, FDI and technology transfer. This also includes certificate training programmes targeting career professionals in the public or private sectors related to foreign direct investment.

- *Researchers* and scientists active in the field of economic development, international economics, international management, foreign direct investment, technology management and technology transfer.

As the target audience of this book primarily consists of managers and students, we have focused on providing concepts and methods applicable in practice. Although theoretical considerations are presented here, our aim has been to keep the presentation of theories and the state of literature relatively short. Also, we have sought to follow an interdisciplinary approach on the topic, integrating concepts and methods from economics on the one hand and the business and management literature on the other hand.

We especially want to thank DIAS (the Direct Investment Advisory Services) and the International Institute of Investment Promotion in Switzerland for supporting this book project. In addition, we want to thank the students participating in the module Technology Transfer and FDI of the M.Sc. Investment Promotion and Economic Development Program offered by Edinburgh Napier University and DIAS for their input to this book.

In order to further advance this book, we are happy to hear comments, critical remarks and feedback.

Boris Ricken and George Malcotsis
Zürich and Baden, Switzerland, April 2011

List of Abbreviations

APCTT	Asian and Pacific Centre for Transfer of Technology
APEC	Asia-Pacific Economic Cooperation
BATNA	best alternative to a negotiated agreement
BOI	Board of Investment (Thailand)
CAGR	compound annual growth rate
CAPEX	capital expenditure
CAST	Chinese Academy of Space Technology
CBERS	China–Brazil Earth Resources Satellite Programme
CORFO	Chilean Economic Development Agency
CRM	customer relationship management
CSIR	Council of Scientific and Industrial Research (India)
DCF	discounted cash flow
DIAS	Direct Investment Advisory Services
DST	Department of Science and Technology (South Africa)
EBRD	European Bank for Reconstruction and Development
EU	European Union
FDI	foreign direct investment
GATT	General Agreement on Tariffs and Trade
GDP	gross domestic product
GNP	gross national product
HICOM	Heavy Industries Corporation of Malaysia
IASP	International Association of Science Parks
ICAMT	International Centre for the Advancement of Manufacturing Technology (Bangalore)
ICSID	International Center for Settlement of Investment Disputes
ICT	information and communication technology
IDA	Industrial Development Agency (Ireland)
IPA	Investment Promotion Agency
IPR	intellectual property rights
IMF	International Monetary Funds
JV	joint venture
MNC	multinational corporation
NESDP	National Economic and Social Development Plan (Thailand)
NIPE	National Institute for Space Research (Brazil)

OECD	Organisation for Economic Co-operation and Development
pa	per annum
PPP	purchasing power parity
R&D	research and development
SEB	Skandinaviska Enskilda Banken
SOFI	Swiss Organisation for Facilitating Investments
SWOT	strengths, weaknesses, opportunities, threats
TRIPS	trade-related aspects of intellectual property rights
UNCTAD	United Nations Conference on Trade and Development
UNESCAP	United Nations Economic and Social Commission for Asia and the Pacific
UNIDO	United Nations Industrial Development Organization
WAIPA	World Association of Investment Promotion Agencies
WEF	World Economic Forum
WTO	World Trade Organization
WWEA	World Wind Energy Association

1

Introduction

1.1 The Role of Technology

Technological innovations have changed the way we experience our private and business life. Physical innovations such as mobile communication technology allow us to connect directly with people around the world at any given time. The Internet facilitates the search and finding of information within seconds. Credit cards and online payment technologies facilitate the buying and payment of goods and services, independently of currency or country. High-speed trains reduce travelling times between metropolitan cities, increasing mobility and flexibility. Medical innovations have resulted in better treatment of diseases or even their prevention. Innovations in seeds and fertilizer technology allow us to grow more resistant, less water-consuming fruits and plants.

New 'soft' technologies have changed the way goods are developed, marketed and sold. Just-in-time production has reduced the required capital stock for goods such as mobile computers or automobiles, making them more affordable for consumers. Innovations in logistics have diminished delivery times and improved the flow of products. Management technologies such as total quality management or business process optimization have reduced production costs and increased the productivity of companies. New marketing techniques have provided a better understanding of consumer's needs and increased client satisfaction.

However, not all new technologies have had a beneficial impact. The rapid and constant technological progress many countries have experienced has caused physical stress, the constant need to multitask and the inability to disconnect from work. Some technologies have also had adverse side effects such as pollution, destruction of environment, negative impact on health, or consumption of non-renewable natural resources.

Nevertheless, new technologies have overall contributed to increased living standards and economic prosperity. Life expectancy, average income and economic wealth have increased. New products and services have been developed, satisfying people's needs in ways that were not met before.

Beyond its relevance for everyday life, technology is of crucial importance for the competitive advantage of firms and nations.

Firms create competitive advantage by inventing new products or finding new technologies to produce and market these. Access to abundant factors, such as cheap labour or resources, is becoming less essential than the technology and skills to produce them effectively or efficiently (Porter 1990: 14). For example, in the first decade of the twenty-first century, American automobile manufacturers have lost competitiveness not because of an unavailability of labour, equipment, or material resources. Rather their product technology, resulting in high fuel consumption, is falling behind European and Asian manufacturers, requiring them to slash prices and cut production. Amongst other factors, this has eventually resulted in the bankruptcy of several large US automotive manufacturers.

For *nations*, assuming that the principal economic goal is to produce a high standard of living, the only meaningful concept of competitiveness is national productivity (Porter 1990: 6; WEF 2009: 4). New physical or managerial technologies provide a competitive advantage by increasing productivity, setting the foundation for a higher standard of living. Factors such as cheap labour, investment in equipment and machinery, the availability of natural resources, low interest rates, favourable exchange rates, or low budget deficits by themselves do not sufficiently explain the long-term growth rates of successful economies (UNCTAD 2005a: 201). For example, nations such as Germany, Switzerland or Sweden have enjoyed rapidly rising living standards despite high wages, long periods of labour shortage and appreciating currencies. Countries such as Japan and Korea have prospered despite budget deficits (Porter 1990: 3). Economies such as Spain and Ireland have to import natural resources and have nevertheless outperformed nations with rich reserves in minerals, oil or gas.

Instead, technological advancements in the field of products and processes have driven long-term economic development. For example, Thailand has attracted manufacturing technology for its electronics industry, strongly contributing to its economic growth over the last decades (see Section 12.3 for details). The economic success of Korea has developed in line with its technological advancement, as measured for example by the number of patents registered (see Figure 1.2). Singapore has also been extremely successful in advancing its economic growth via the acquisition of new technologies in the field of electronics, biotechnology and banking (Box 1.1). Ireland, also referred to as 'The Celtic Tiger', has experienced substantial economic progress through FDI and advancements in its software and biotechnology industries (see Section 15.7).

Few if any countries have succeeded in achieving and sustaining high growth levels without investing in and exploiting new technologies (UNCTAD 2005a: 201). The importance of technology for economic development can therefore hardly be overstated.

BOX 1.1   SINGAPORE – A SUCCESS STORY OF TECHNOLOGY TRANSFER

With a population of about five million (2009), Singapore has shown impressive economic growth over the last years. Since 1980, real gross domestic product (GDP) has grown on average at 6.4 per cent per year, resulting in an increase of real GDP by close to six times. How could a small economy with no natural resources and domestic market achieve such a success?

A major driver has been technology transfer. Singapore went through four different phases of technological evolution (see UNCTAD 2003a: 50):

- **Industrial take-off** (early 1960s–mid 1970s). In this first phase, Singapore showed a high dependence on technology transfer from multinational corporations (MNC).
- **Local technological deepening** (mid 1970s–late1980s). The second phase was characterized by rapid growth of local process technological development within MNCs and the development of local supporting industries (clusters).
- **Applied R&D (research and development) expansion** (late 1980s–late 1990s). Later, a rapid expansion of applied R&D by MNC, public R&D institutions and local firms took place.
- **High-tech entrepreneurship** (late 1990s onwards). The last phase was characterized by the emphasis on high-tech start-ups and the shift towards technology-creation capabilities.

> In these four phases, Singapore shifted from emphasizing technology *use* to technology *creation*, and each phase built upon resources accumulated in earlier phases. Skill creation, advanced infrastructure, strategic policy-making and efficient administration were key success factors in climbing up the technological ladder. Despite a free trade stance to attract FDI, the Singapore government was highly interventionist, deploying a battery of selective measures to support the technological evolution of the country. As a result of these measures, the country today ranks third in the world with regard to its global competitiveness.
>
> (WEF 2009: 276).

However, the gap in technological advancement between countries is large. For example, expenditure on R&D formulates a key indicator of government and private sector efforts to obtain competitive advantage in science and technology. Figure 1.1 illustrates the percentage of GDP large countries spent on R&D in 2007. Korea is the only newly industrialized economy among the top 15 spenders on R&D and China is the only developing economy among the top 20. With 3.4 per cent the industrialized nation Japan spent a more than seven times higher ratio on R&D than the developing country Mexico, with its 0.46 per cent of GDP. Given that the Japanese GDP exceeds that of Mexico by three times (valued at purchasing power parity – PPP), the gap is even wider.

Similar observations can be made for the share countries hold in world's registered patents (see Figure 1.2). Patents provide a measure of the output of a country's R&D, that is its inventions. The top three countries, the US, Japan and Germany account for over 70 per cent of the world's patents. There was not one emerging economy among the top 10 patent takers in 2006.

In order to bridge this technology gap, foreign sources of technology are of dominant importance for developing countries (Eaton and Kortum 1999). Channels to transfer technologies from these foreign sources are trade in products, trade in knowledge, or international movement of people, but one of the most important channels is foreign direct investment (FDI). Many economies seek FDI not only to attract finance capital: they are also interested in technology spillovers to domestic firms.

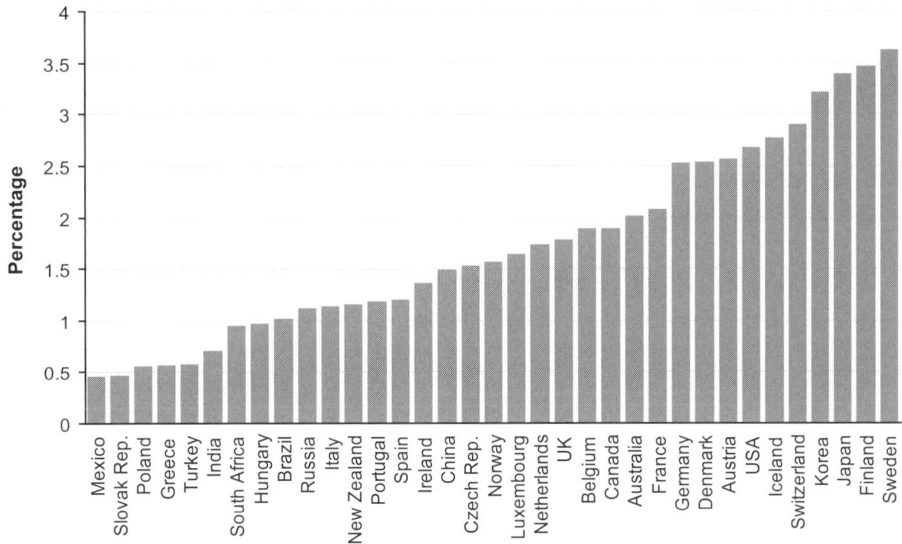

Figure 1.1   Gross domestic expenditure on R&D in 2007 as a percentage of GDP

Data source: OECD.

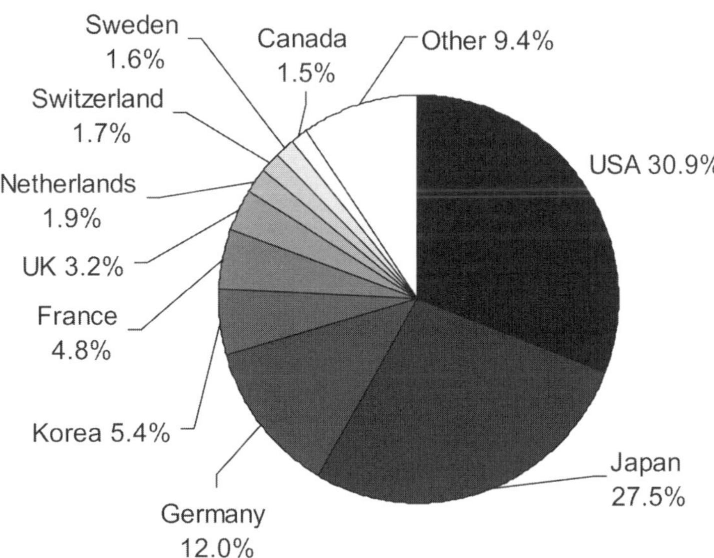

Figure 1.2   Share of countries in triadic patent families in 2006

Data source: UNCTAD. Triadic patent families are sets of patents taken at the European Patent Office, the Japanese Patent Office and the US Patent and Trademark Office.

1.2 The Aim of this Book

The importance of technology transfer via FDI implies that countries cannot simply sit and wait until new technologies arrive in their domestic domain. Rather, companies, regions and nations need to systematically manage the identification, attraction and absorption of new technologies.

This book wants to provide a *practical guidance* for companies (joint venture partners, local competitors, suppliers, buyers), government bodies (investment promotion agencies, investment boards, ministries of economics or ministries of science and technology), and multilateral organizations (development banks and agencies) on how to effectively attract, absorb and retain new technologies and knowledge via FDI. It aims to provide a step by step guide on how to manager the entire technology transfer process in FDI, starting with the development of a technology strategy, continuing with the assessment of technologies, their attraction, absorption and commercialization and ending with the controlling of the achieved technological progress. This book wants to support managers in answering the following key questions:

- What technologies should our region or country focus on?

- How can we get information on these technologies and assess their benefits, costs and unwanted effects?

- What can we do to attract investments containing these technologies?

- How can we facilitate the absorption of these technologies by our local companies?

- How can the companies in our country commercialize and use these technologies?

- How can we evaluate if we were successful in our efforts?

1.3 The Structure of this Book

In order to achieve this aim, we have divided the book into two parts (see Figure 1.3).

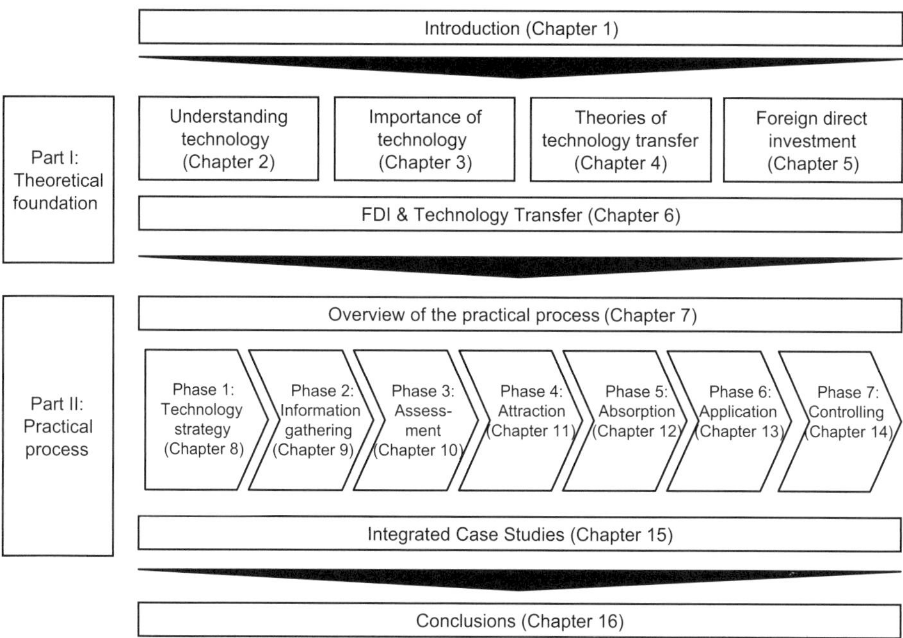

Figure 1.3   The structure of this book

Part I (Chapters 2–6) provides the *conceptual and theoretical foundations* on technology transfer and FDI. We base our theoretical discussion on the economic literature, having studied FDI intensively, as well as the managerial literature, having intensively analysed technology and knowledge transfer:

- Following this introduction, Chapter 2 develops a clear understanding of the term technology, describes characteristics of technology and differentiates between various types of technology. In addition, it illustrates several mediums for technology storage and the various levels of technological capabilities a region or nation may posses.

- In Chapter 3, we deepen the understanding of technology by describing its importance for companies and nations with the help of managerial and economic theories.

- Chapter 4 lays out three different theories of technology transfer. These allow us to produce generic key factors for technology transfer depending on the type of technology to be transferred.

- Following this, in Chapter 5 we develop an understanding of FDI by describing historic and recent trends in this area, defining the term and identifying different forms of FDI. We then look at determinants of a firm's investment decisions and the contribution of FDI to the local economy.

- Finally, Chapter 6 sets technology transfer in the context of foreign direct investment. We distinguish between three different channels of technology transfer in FDI: (1) an *internal transfer* of technology between a MNC of a home country and its local subsidiaries in a host country; (2) an external transfer via a *horizontal spillover* from the local subsidiary to competitors; and (3) an external transfer via a *vertical spillover* from the local subsidiary to suppliers or buyers in the host country.

In Part II we give an overview on the *practical process of managing technology transfer* in FDI (Chapter 7). We divide this process into the following components (see Figure 1.3):

- The process starts with the *formulation of a technology strategy* (Chapter 8). This includes the development of a technology mission statement, a strategic analysis, the formulation of strategic themes and a raster for the implementation of the technology strategy.

- In the second phase, managers and investment promotion agencies need to *gather information* on technologies of interest to them (Chapter 9). We show you what information categories are relevant, from which sources information can be gathered and which searching techniques can be applied.

- In the third phase, managers should *assess and evaluate* the technologies in question (Chapter 10). Benefits, costs and unintended consequences of introducing certain technologies into a country need to be assessed. A variety of methods such as trend-line analysis, modelling, scenarios, Delphi survey, discounted cash flow (DCF) or cross-impact analysis help you in doing so.

- In a next step, a selected technology needs to be *attracted* to a host country (Chapter 11). To do so, managers first require knowledge on how to attract FDI to their country. Secondly, managers need to understand policies facilitating the transfer of technology to the local subsidiary of these FDI projects.

- Simply attracting a technology to a country or region does not yet guarantee that the technology is *absorbed* by local companies. For this to take place, either horizontal or vertical spillovers have to occur. We therefore provide you with a battery of policies to raise the absorptive capacity of a country and to facilitate interaction between the foreign investor and local companies (Chapter 12).

- The *application* of technology in a firm, industry and country formulates the next phase of active technology management (Chapter 13). While acquisition and absorption are necessary preconditions for this step, they do not yet guarantee the actual use of the acquired technology. In the case of process technologies, we show how internal motivational, technological and organizational barriers can be overcome. For product technologies, we show how a market strategy is developed and implemented.

- The last phase in the technology management process consists of *performance measurement* (Chapter 14). Performance measurement in technology transfer allows managers to measure the achieved progress and to learn from experience. We will provide you with the concepts and tools to develop a performance measurement for the entire process of managing technology transfer.

Afterwards, we will illustrate the relevance of the various aspects of the process of managing technology transfer via six integrated, in-depth case studies, covering different countries and industries (Chapter 15). We finish this book with some conclusions (Chapter 16).

Theoretical Foundations

2

Understanding Technology

We start this book by providing you with a clear understanding of technology. To do so, we structure this chapter as follows:

- First, we give you a definition of knowledge and technology by setting both terms in the context of the related concepts 'symbol', 'data' and 'information' (Section 2.1).

- Subsequently, we describe various characteristics through which technology can be distinguished from other production factors such as labour or resources (Section 2.2).

- Then we explain different types of technology, distinguishing between hard and soft, explicit and tacit, and process and product technology (Section 2.3).

- In addition, we illustrate various media through which technology can be stored (Section 2.4).

- Following this, we differentiate between five levels of technological capabilities a country may possess (Section 2.5).

- Finally, we advance a dynamic view of technology, describing various patterns of technological evolution (Section 2.6).

2.1 Defining Knowledge and Technology

Any discussion of technology transfer in FDIs requires a definition of technology in the first place. As technology and knowledge are closely related to each other and both terms are often used interchangeably, we first want to examine the term knowledge before defining technology.

2.1.1 KNOWLEDGE

In theory as well as practice the term knowledge is not used in a coherent way and is subject to various interpretations. Defining knowledge has occupied philosophers for thousands of years (Holsapple 2003: 165). The consideration of knowledge presented in this section is comparably modest in scope but aims to capture the basic ideas required for further analysis. In order to develop a clear definition we define knowledge as a concept that is founded on several hierarchical layers (see Figure 2.1, Rehäuser and Krcmar 1996: 5, Sanchez 2001: 5, Mutch 2008: 41–65):

- A symbol is the most basic element. Symbols may be either letters or numbers. They are not related to each other in any way. Examples for symbols are 'A', 'd', or '3'.

- Data is created through the combination of various symbols. Symbols become data when they are organized through a code such as the grammar of language or mathematical definitions. Data can be useful or not useful but of itself data has no meaning. Its usefulness and interpretation depend on the context. For example, take one piece of data: NJ630117. This could be a stock code or a customer reference number. It is in fact a British Grid reference. Using a map we can find a certain location with this data (see Mutch 2008: 459). Also, the symbols 'r', 'e', and 'd' form the data element 'red'. 'Red' makes a declaration about something but at this level of the knowledge hierarchy the context remains unclear.

- Information is data organized in a problem context. Data becomes information if it is positioned in a context. Information is the meaning that is imputed to some data by evaluating it in an interpretative framework. An example of information is the sentence 'This car is red'.

- Knowledge is a set of beliefs about causal relationships in the world. Information becomes knowledge when its holder develops an understanding of the relations between sets of information. The concept of knowledge makes a distinction between simply being aware of something, which means having data or information, and having knowledge, which implies actually knowing how to do things or cause things to happen. An example of common knowledge is expressed in the sentence 'If I let this apple fall, it will hit the ground'.

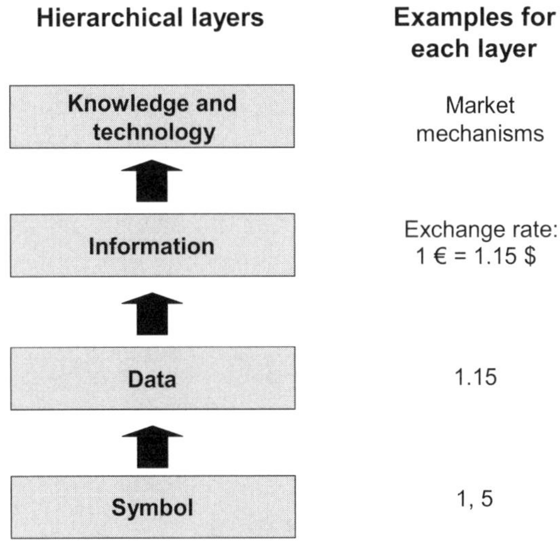

Figure 2.1   A knowledge hierarchy

BOX 2.1   A PRACTICAL EXAMPLE TO ILLUSTRATE THE KNOWLEDGE HIERARCHY

In the following, we use the example of an investment broker to illustrate the differences between symbols, data, information and knowledge. A broker in Frankfurt receives the information that the actual exchange rate of the euro to the US dollar is 1.15 €/US$. This information is made up of symbols such as '1', ',', '1' and '5' and data elements such as '1.15', 'US$/€'. The broker also receives information on a rise in the US interest rate. The broker has some knowledge of market mechanisms allowing him to establish relations between the received information and to make an investment decision. This knowledge consists of the belief that a rise in the US interest rate will attract additional foreign capital to the US, creating a growing demand for the dollar. Based on this knowledge and the market information the broker buys US dollars, speculating on a rise in the value of the dollar.

Adapted from Rehäuser and Krcmar (1996: 5).

2.1.2 TECHNOLOGY

Subsequently to knowledge, we define the term technology. Technology is a word with origins in the Greek word technologia ($\tau\epsilon\chi\nu o\lambda o\gamma\iota\alpha$), consisting of techne ($\tau\epsilon\chi\nu\eta$) 'craft' and logia ($\lambda o\gamma\iota\alpha$) 'saying'. In this book, we follow an encompassing understanding of technology. Technology does not only refer to hard technology, such as machines or patents, but also to soft technology, such as management expertise and professional know-how (see section 2.3.1 for details). Technology may be defined as the knowledge of how to perform tasks, solve problems and provide products and services in organizations. Technology denotes not only the sum of knowledge, experience and skills necessary to manufacture and market a product economically: it also refers to the knowledge necessary for the planning, establishment and operation of a firm (UNCTAD 2001: 6). In the context of a firm, we therefore use the terms technology and knowledge interchangeably (see also Jindra 2006: 8, Figure 2.1).

2.2 Basic Characteristics of Technology

Technology heavily influences the production process of goods and the provision of services. However, technology may be distinguished from other inputs through a variety of characteristics (Rehäuser and Krcmar 1996: 10):

- *Ownership*: other than traditional factors of production technology can be owned by a variety of persons. While machinery and labour exist in materialized form and are only present at one place at a time, the same technology may be applied at different locations.

- *Usage*: traditional factors of production wear and tear in the production processes lose value. Technology does not show such effects. Nevertheless, it may lose market value through its distribution to others.

- *Reproduction*: the reproduction of traditional factors of production generally causes considerable costs. On the other hand, the reproduction of technology generally causes few costs with regard to storage mediums, for example CDs, books, or magazines. However, costs may arise through the transfer of personalized knowledge via training etc.

- *Price*: the prices of traditional inputs such as labour or machinery are determined on national or international markets. Technological knowledge, on the other hand, usually lacks a market because of its special characteristics. To judge the value of a technology, the buyer has to fully understand its content and functions. Once he can do this, no incentives remain to pay for the technology. The seller of technological knowledge will therefore never completely reveal the characteristics of his technology. This is why formulating a price for technology is extremely difficult and always subject to negotiations.

- *Protection*: machinery or products may easily be protected by the owner through a serial number or physical enclosure. Despite this, technology may be distributed across physical barriers, making its protection more difficult for the owner.

2.3 Types of Technology

For the transfer of technology in FDI, it is important to know which type of technology will be transferred. The managerial literature has developed a large number of typologies of knowledge and technology. For the purpose of this book, we want to differentiate between three dimensions:

- Hard vs. soft technology (Section 2.3.1);

- Explicit vs. tacit technology (Section 2.3.2); and

- Product vs. process technology (Section 2.3.3).

2.3.1 HARD AND SOFT TECHNOLOGY

First, we can distinguish between hard and soft technology (see Dyker 1999: 9):

- Hard technology is the knowledge necessary to carry on or to improve the existing production and distribution of goods and services.

- Soft technology is entrepreneurial expertise and professional know-how, as in the fields of management, organization or marketing.

The effective assimilation of industrial technology is not just a matter of mastering the use and development of hard technology. The soft dimension of technology is also critically important. For instance, the introduction of new production processes into any organization will usually require some adaptive change in the organization and the procedures of production. Box 2.2 illustrates the state of hard and soft technology in the former Soviet Union at the beginning of the 1990s.

BOX 2.2   HARD AND SOFT TECHNOLOGY IN THE FORMER SOVIET UNION

Investors entering the former Soviet Union observed that the hard technology of some industries was fairly satisfactory. There were cases where factories had been taken over by foreign companies and switched to exporting for hard currency, without any major changes being introduced on the actual production floor. Why then was the economic performance of so many firms unsatisfactory?

Much of it can be explained by the shortfalls of soft technology. For example, under the communist system investment planning and strategic issues were particularly weak and overridden by bureaucratic goals. Also, the notion of design as the factor integrating innovation, cost-effectiveness and marketing was barely grasped. This lack of soft technology was to a large extent responsible for low productivity and poor performance.

Adapted from Dyker (1999: 9).

2.3.2 EXPLICIT AND TACIT TECHNOLOGY

For the transfer of technology it is not only critical if technology is hard or soft. The degrees of explicitness and of codification are also fundamental characteristics of technology. The explicitness of knowledge is the extent to which an individual has articulated some knowledge in an understandable form. Along this dimension, we can differentiate between explicit and tacit forms of technology (Polanyi 1966):

- *Explicit technology* can be articulated by individuals in written or verbal form. It can be codified, that is, placed in some ordering framework or structure. It is to some degree abstract. Explicit technology may be transferred via media such as language, books, or documents. It may also be in the form of audio, vision, or other forms of multimedia representations. An example of explicit technology is the documentation of the organizing process of a typical investment promotion event. Technological knowledge that can be expressed in words and numbers only represents the tip of the technology iceberg (see Figure 2.2).

- *Tacit technology* is the implicit knowledge used by organizational members to perform their work and to make sense of their world. It expresses the idea that we know more than we can tell. Tacit technology is hard to verbalize because it is expressed through action-based skills and cannot be reduced to rules or recipes. It is highly personal, context-specific and deeply rooted in individual experiences, values, ideas and emotions. Nevertheless, tacit technology underlies almost any action and decision. It therefore formulates the bottom of the technology iceberg (see Figure 2.2). Common examples of tacit technology include the knowledge of how to ride a bicycle, play the piano, or how to use a word processor. The transfer of tacit technology is usually slow and can often only occur to a limited audience.

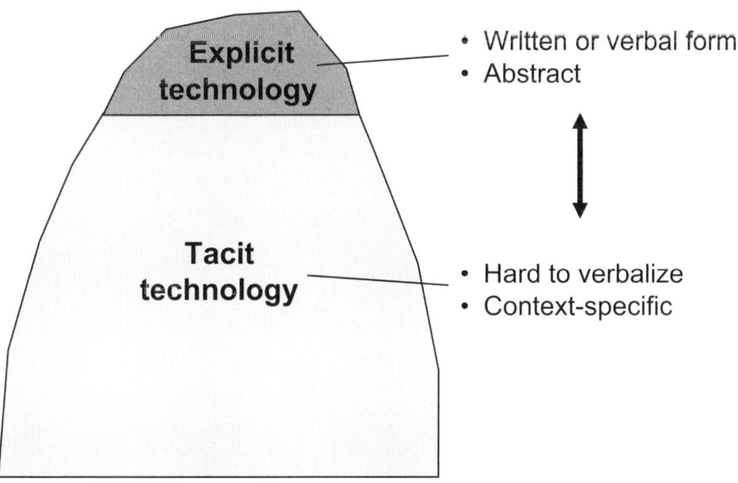

Figure 2.2   The technology iceberg

We will examine the importance of tacit and explicit technology for technology transfer in more detail in Chapter 4.

2.3.3 PRODUCT AND PROCESS TECHNOLOGY

As a third typology, we distinguish between product and process technology (see Adner and Levinthal 2001, Bogers 2009):

- *Product technology* refers to the knowledge related to a particular product a firm is selling for the benefit of customers or clients. A change in product technology may either occur through the production of a new product, such as a new type of medication in case of the pharmaceutical industry, or it may involve a new technology in already existing products, such as using hybrid engines in automobiles. Large firms are usually in possession of specialized R&D functions, explicitly dedicated to generating innovations in product technologies.

- *Process technology* refers to the tools, materials, equipments and other technologies that directly affect how a firm develops, produces, or distributes the goods and services it sells on the market. Changes in process technology can range from minor improvements to radically new technologies. Even incremental innovations in process technology can lead to significant benefits. For example, it has been shown in a study for DuPont Rayon Plants that about 80 per cent of cost reductions in the plants were derived from minor technical changes (Hollander 1965). In a similar study on digital computers, it turned out that performance advances were the result of small improvements by equipment designers (Knight 1963). A typical characteristic of process technology is therefore that it usually does not depend on formal R&D efforts but rather is developed through informal activities such as 'learning by doing' (Bogers 2009: 9).

During the evolution of an industry or product life cycle, the relative importance of product and process technology may change. A firm might first gain a differentiation advantage by launching a new and innovative product. Once its competitors have been able to copy this innovation, the same company might achieve a cost differentiation by enhancing its process technology and producing at substantially lower costs than other companies.

2.4 Mediums for Technology Storage

Depending on its characteristics technology may be stored in different mediums (Al-Laham 2003: 36–42). We differentiate between hard, human and collective mediums.

Hard mediums focus on technology that can be articulated in verbal or written form. Technology may be stored in hard mediums for functions such as:

- documentation of technology (protocols, process documentation);

- duplication of technology (publications); and

- transfer of technology to other persons (via teaching material, direction of use).

Some hard mediums that may be used to store technology are:

- Print-based mediums such as books, journals or written notes. For example, many companies use internal magazines in order to spread and promote the usage of certain process technologies such as software programs or certain accounting procedures.

- Audiovisual mediums like audiotapes, photos, or videos. For instance, some firms hold introduction courses for new employees which often make use of videos delivering certain knowledge about the company.

- Computed-based mediums such as the Internet, intranet, CD, DVD, hard disc. For example, many companies have developed internal knowledge databases that employees may access via an intranet.

- Product-based mediums such as assembly lines, production factors, or finished products.

Human mediums consist of the individual members of a firm. Other than technical systems humans learn out of their experiences and are also capable of applying technology. Different from computers, organizational members do not store technology in a representational form. Instead technology is interpreted in the light of prior knowledge, values and beliefs: therefore the technology stored by individuals is always subjective. Individuals may accumulate explicit as well as tacit technology.

Collective mediums refer to the observation that technology may not only be stored within the minds of single individuals. Groups or organizations may also possess a common stock of technological knowledge. Such a stock is shared by all members of an organization and is to a large extent independent of individuals. Individual changes in a firm do not affect the collective stock of technology in a company such as organizational routines, processes or culture.

2.5 Technological Capabilities

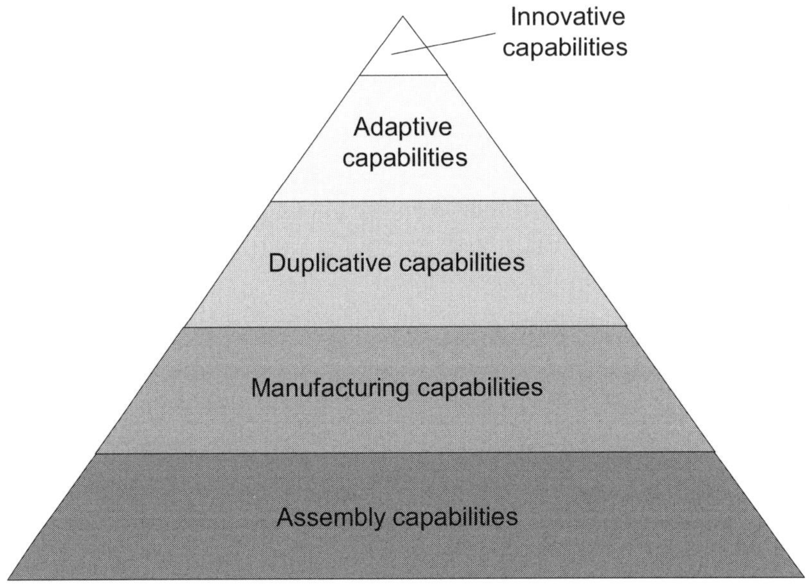

Figure 2.3   A pyramid of technological capabilities

So far we have seen the various characteristics of technology, its different types and the distinct mediums through which technology can be stored. In the context of FDI, it is also essential which level of technological capabilities a region or country holds. We can differentiate between five levels of technological capabilities (see Figure 2.3, see Lall 2003: 22, UNCTAD 2005a):

- At the bottom level are assembly capabilities. Assembling a final product is capital- and labour-intensive and requires lower technologies than other stages of production. More demanding skills such as troubleshooting, quality control, maintenance and procurement skills are included at this level.

- At the second level we find manufacturing capabilities. The manufacturing of components requires capital-intensive investment for mass production. Reaching internationally acceptable levels of production efficiency and quality in complex activities is very demanding. Many enterprises fail to do this, even after years of operation, unless they invest sufficiently in collecting information, creating new skills and developing appropriate management structures.

- At the intermediate level are duplicative capabilities. These include the skills needed to expand capacity and to purchase and integrate foreign technology extensively. Licensing an existing technology or reverse engineering activities fall into this level (see Box 2.3 for an example). A typical example of duplicative capabilities is a fast-food store that uses licensed product and marketing technology from its franchise partner in order to produce and sell its products.

- The second-highest level consists of adaptive capabilities. These allow adaptation and improvement of imported technologies. The adaptation of a technology is particularly challenging if conditions are significantly different from those at the origin of the technology and if local support and supply structures are weak.

- Finally innovative skills, based on formal R&D, are needed to keep pace with technological frontiers and to generate new technologies. Here firms design, develop and test entirely new products and processes.

Throughout their technological development, countries and regions improve their capabilities and move up the pyramid. It is noteworthy that these capabilities each focus on a certain industry. A country may have reached innovative capabilities for one industry but for another it may only have reached the level of duplicative skills. For example, India's software industry has reached the level of innovative skills, being capable of pursing own R&D

projects and generating new software technologies. However, other industries in India, such as the biotechnology or machinery industry have not yet reached this level of technological capability and still rely predominately on duplicative and adaptive skills.

BOX 2.3   REVERSE ENGINEERING IN INDIA'S PHARMACEUTICAL INDUSTRY

Reverse engineering is the process of taking something – for example a consumer product, a machine, or pharmaceuticals – apart and analysing its workings in detail. In the business world this is usually done with the intention of deconstructing the technology stored in a new device to be able to reproduce it.

Before becoming a member of the GATT (General Trade Agreement on Tariffs and Trade) (now the World Trade Organization [WTO]) in 1994, India allowed only process patents for pharmaceuticals. These permitted the creation of the same product through a different process. Indian pharmaceutical companies took this opportunity to flourish in the healthcare sector. Many Indian companies mastered reverse engineering and produced drugs in India which were patented elsewhere in the world. Indian pharmaceutical companies underwent strong growth and achieved the status of providing quality medicines at affordable prices.

2.6 Technological Evolution

So far we have adopted a static view of technology, but technologies evolve over time. Managing technology transfer therefore requires some understanding on the evolution of technologies, allowing users to anticipate changes in technologies or in their usage. We introduce two different models of technological evolution.

First, the technology S-curve is a graphical representation of the development of a new technology. The S-curve typically compares some measure of performance (for example, speed, cost, or capacity) with some measure of effort (R&D investment, or hours devoted to research). The graph is called an S-curve because the relationship between effort and performance is typically S-shaped (Shane 2009: 22, see Figure 2.4):

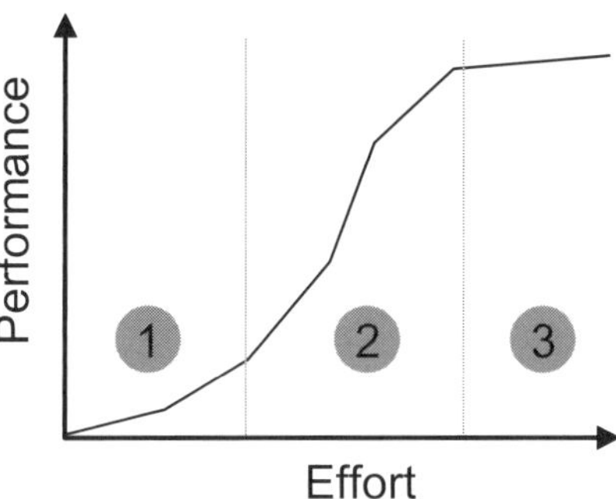

Figure 2.4   The technology S-curve

1.   Initially, performance improvements per unit of effort are small because the key drivers of performance are not yet well understood. This phase involves much trial and error and many dead ends.

2.   Once the key drivers of performance are identified, the rate of performance improvement for each unit of effort increases. For example, the first airplanes required a significant investment of research on wing and engine design to achieve very short flights, but once the basic design of the airplane was identified, the performance of aircraft increased strongly.

3.   However, at some point the technology reaches its physical limits and diminishing returns begin to set in, leading to lower additions in performance for each unit of effort. For example, the fuel performance of automobiles has not been dramatically improved in recent decades although significant investments in R&D have been undertaken by the large automobile manufacturers.

While the S-curve relates performance and effort to each other, one may also relate variations in technology to the evolution of a technology. Empirical studies typically identify three phases (see Agarwal and Tripsas 2008: 6–13 for an overview):

1.   *Innovative activity.* In the nascent stage of a technology, technical variety is high and a diverse range of innovations exists. Technologies look different, emphasize different functions and offer different features. For example, in early automobiles, steam and electric

engines competed with the eventually dominant combustion engine. Also, the round steering wheel competed with a joystick-type tiller for controlling the direction of the vehicle (Albernathy 1978). Another example is the early radio transmitter which used alternator, arc and vacuum tube technologies before vacuum tubes became dominant (Aitken 1985).

2. *Emergence of a dominant design.* A key turning point for the evolution of many industries is the emergence of a dominant design. A dominant design refers to standardized modules and interfaces incorporating a particular set of features. For example, in the automotive industry, such a dominant design has eventually emerged with the combustion engine and the closed body vehicle. Another technology which eventually developed a dominant design is the personal computer. Once such a dominant design has emerged, firms focus on improving and extending it, resulting in periods of incremental technological change. The emergence of a dominant design is often accompanied by an accelerated diffusion of a technology.

3. *Technological discontinuity.* Following the establishment of a dominant design, it can happen that a new period of turbulence occurs and the dominant design is challenged again. Such rivalries existed between the diesel locomotive vs. the steam locomotive, the jet engine vs. the propeller or the ballpoint pen vs. fountain pens (Cooper and Schendel 1976).

3

The Importance of Technology

In this chapter, we want to deepen our understanding of technology and provide some insight as to why technology is important for firms and countries. We illustrate the significance of technologies through two different theoretical streams:

- First, we show the relevance of technology for the competitive advantage of companies with the help of strategic management literature (Section 3.1).

- Secondly, we point out the importance of technology for the economic growth of nations by explaining its role in macroeconomic growth theory (Section 3.2).

3.1 Firms: Generating a Competitive Advantage

A firm can be understood as a collection of activities to design, produce or market its products and services (Porter 1985:36):

- *Primary activities* are those activities involved in the physical creation of a product, its sale, transfer to the buyer and then after-sales assistance. In a nutrition company producing cereal, for instance, the production of the cereal would be part of its primary activity.

- *Support activities* support the primary activities and each other by providing purchased inputs, technology, human resources and various company-wide functions. Typical supporting activities in a bank, for example, are recruiting and training of employees, internal IT support or procurement of office material.

Technology is embodied in every single one of these activities. 'Everything a firm does involves technology of some sort, despite the fact that one or more technologies may appear to dominate the product or the production process' (Porter 1985: 166). Examples for technologies contained in a firm's *primary activities* (see also Porter 1985: 167) are:

- inbound logistics: transport technology, materials handling technology, etc.

- operations manufacturing: basic process technology, machine tool technology, production process technology, etc.

- outbound logistics: transportation technology, packaging technology, logistics optimization technology, etc.

- marketing and sales: media technology, Web technology, customer relationship management (CRM) technology, marketing technologies, etc.

- service/after-sales assistance: diagnostic technology, testing technology, etc.

Typical technologies involved in a firm's *support activities* (see also Porter 1985: 167) are:

- human resource management: employee assessment technologies, motivation research, training technologies, compensation models, etc.

- procurement: demand forecasting technologies, information system technology, etc.

- technology development: R&D process design, creativity techniques, product development technologies, computer-assisted design, etc.

- firm infrastructure: planning and budgeting technology, office technology, etc.

Every value activity of a firm therefore uses some technology to combine purchased inputs and human resources to produce some output (Porter 1985: 166).

As activities and technologies in firms are linked to each other, a technological choice in one part of the value chain can have implications to other parts of the chain. Thus a firm cannot only be understood as a collection of activities but also as a chain of linked technologies (Porter 1985: 166–169).

Technology and knowledge can therefore be described as strategically significant resources of a firm (Grant 1996: 375) – they are a strategically important resource because they are scarce and not easily transferable or replicable (Grant 1996: 375). This accounts especially for tacit technology, which is hard to transfer across organization boundaries. Because the firm's value activities all represent different types of specialized technology, the competitive advantage of a company depends on the way a firm is able to integrate these technologies. For example, a hospital's capability in cardiovascular surgery is dependent upon integrating the specialist knowledge of surgeons, the anaesthetist, radiologist, operating room nurses and several types of technicians (Grant 1996: 377).

The combination of its technologies may allow a firm to gain a *cost advantage* or a *differentiation advantage* (Porter 1985: 169):

- A firm may gain a *cost advantage* by using a technology to produce at lower cost than its competitors. An example of this is the Japanese cement industry. Due to increasing energy costs, Japanese cement producers became less and less cost competitive in their core export markets in Asia. Through the use of waste as an alternative fuel to fire their cement kilns, companies managed to reduce their production cost, regaining competitive advantage over other Asian cement exporters from Taiwan and Korea.

- A firm may also gain a *differentiation advantage* over its competitors by applying a new technology. The introduction of the hybrid engine for automobiles is an example for such an advantage. Hybrid engines generate value for buyers of automobiles because they consume less petroleum than conventional engines, thereby running at a lower cost. Japanese firms, which first introduced hybrid engines into the market, enjoyed a unique position in the industry due to this new technology.

Technological change is therefore one of the principal drivers of competition (see Box 3.1 and Box 3.2 for examples).

BOX 3.1   THE COST ADVANTAGES OF DELL'S SUPPLY CHAIN

A typical example for technology change creating cost advantages is the strategy pursued by Dell Computers. Prior to Dell personal computers were manufactured in volume, shipped to retail stores and sold individually to customers. This required massive amounts of inventory, and customers were limited to a relatively small set of configurations. Dell applied new technologies in their value chain by adopting a direct sales strategy, building every PC to order and shipping it directly to the customer. Technological innovation in the value chain allowing direct sales and just-in-time production reduced storage costs considerably. It even gave the company the financial advantage of a negative cash-to-cash time: the company was paid for its products before it bought the components. It is estimated that the Dell direct model created a 4–6 percentage point margin advantage over historical indirect channel models.

Adapted from Taylor (2010).

BOX 3.2   THE DIFFERENTIATION ADVANTAGES OF FEDERAL EXPRESS

Federal Express (FedEx) modified the technology of its outbound logistics for small parcel delivery. The company, which moves about 7.5 million packages around the world every day, achieved faster and more reliable delivery than other firms. Also, in November 1994 FedEx provided public access to the company's internal package-tracking database. The company was surprised to find that a very high number of customers actually preferred tracking their own packages on the company's website, as opposed to speaking in person to a customer service representative on the phone. Package-tracking via the Internet created additional customer satisfaction and laid the basis for differentiation advantages.

3.2 Nations: Increasing Factor Productivity

Technology does not only shape the competitive advantage of single firms, it also impacts the economic growth and competitive advantage of regions and nations. For example, in 1996 real GDP of the UK was ten times its level of 1870. Over the same period, real GDP in Japan increased 100-fold and real income per person 27-fold (Begg et al. 1997: 503–517). How can such long-term growth rates be achieved? As we will see in Box 3.3, Box 3.4 and the rest of this section, technological progress exerts a critical influence on economic growth.

BOX 3.3   THE PREDICTIONS OF THOMAS MALTHUS

One of the earliest doomsters was Thomas Malthus, who wrote his 'First Essay on Population' in 1798. Living in a largely agricultural society he worried about the fixed supply of land. As the population was growing, Malthus predicted that agricultural output would fail to increase in line with this growth. He predicted that starvation and death would reduce population to the level that could be fed from the given supply of agricultural land.

Yet Malthus' prediction did not prove correct. Most countries managed to break out of the Malthusian trap because of improvements in agricultural productivity. A rapid technical progress in agricultural production led to large and persistent production increases. The replacement of animal power by machines, the development of fertilizer, drainage and irrigation and new hybrid seeds even allowed some of the workforce to be freed to produce industrial goods while maintaining sufficient food production to feed the growing population.

Adapted from Begg et al. (1997: 503–509).

To analyse the importance of technological progress we first need to look at how the output of an economy and economic growth is determined. To organize our ideas, we start with the production function (see Begg et al. 1997: 503–517). The *production function* shows the maximum output an economy can produce using specified quantities of inputs.

A simple production function is:

$$Y = A \times f(K, L)$$

Where Y is the economic output, K the capital and L the labour input. The function f(K, L) tells us how much we get out when we use special quantities of the inputs K and L. Technical progress is separately captured through A, which measures the extent of technology at any date. The production function tells us that increases in potential output can be traced to increases in inputs of the factors of production – labour or capital – or to technical advances allowing the existing factors to produce a higher level of output.

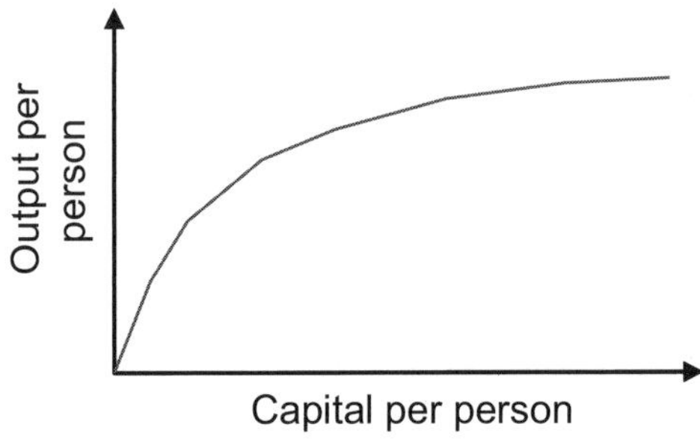

Figure 3.1   Output growth related to capital

A *growth in capital* may be caused by a higher savings rate of a society or the attraction of foreign investment. Growth in capital leads to a higher productivity per person, but with diminishing returns (see Figure 3.1). While one new machine for a worker might raise his output considerably, two or three additional machines will not produce the same effect. That is why economic growth based on capital investment has its limitations (Begg et al. 1997: 503–517).

As the production function indicates, economic output may also increase because of *a rise in the input of labour*. Labour input depends on the number of people working as well as the hours worked. For the last factor, the increase is limited as one can only work a certain amount of time each day. The first factor may be increased with a growing population and may lead to a rise in the overall economic output. However, population growth will not result in an increase of per capita output. Also, population growth typically leads to a set of social, economic and environmental problems related to overpopulation. Its role in long-run growth processes is limited.

Higher accumulations of capital and population growth only have a limited effect on the increase of output (see also Malairaja and Zawdie 2004: 235). While explaining part of the growth of economies, an increasing amount of both factors is not sufficient to explain the tremendous long-term growth rates mentioned at the beginning of this chapter. Technological advances fill this gap by explaining much of modern economic growth in industrialized as well as developing countries (see for example Chenery 1960, Kuznets 1966). Technological advances come through the discovery of new knowledge (*inventions*) and the incorporation of new knowledge into actual production techniques (*innovations*) (Begg et al. 1997: 503–517). In economic history inventions such as electric light, the automobile, or the internet are evident examples for the tremendous growth in technical knowledge in industry. As the production function indicates, an increase in A influences the achieved output considerably. '[F]ew, if any, countries have succeeded in achieving and sustaining high growth levels without investing in and exploiting technology' (UNCTAD 2005a: 201). Box 3.4 illustrates this by describing the role of technology in China's long-term economic growth.

BOX 3.4   CHINA'S ECONOMIC GROWTH AND THE ROLE OF TECHNOLOGY

In 1978 the government of China embarked on a major programme of economic reform. It encouraged the formation of rural enterprises and private businesses, liberalized foreign trade and investment, relaxed state control over some prices and investment in industrial production and the education of its workforce. According to an IMF study, post-1978 China saw an average real growth of more than 9 per cent per year. Per capita income nearly quadrupled between 1982 and 1997.

Capital accumulation such as the growth in factories, manufacturing machinery and communication systems was important for the economic boom, as were the number of Chinese workers. However, despite a huge expenditure of capital, production of goods and services per unit of capital remained about the same. A sharp, sustained increase in labour productivity caused through the use of new management technology was the driving force behind the economic boom.

From 1979–1994 China's productivity gains account for more than 42 per cent of China's growth. By the early 1990s, productivity's share of output growth exceeded 50 per cent, while the share contributed to capital formation fell below 33 per cent. Productivity increase caused by technological progress had overtaken capital as the most significant source of growth.

Adapted from Hu and Khan (1996).

4

Theories of Technology Transfer

In the previous chapters we have defined technology (Chapter 2) and gained a first understanding of the importance of technology for companies and nations (Chapter 3). In this chapter we focus on the aspect of transfer. We introduce three different theories explaining the transfer of technology:

- First, you will become acquainted with *Nonaka's theory of technology conversion* (Section 4.1). This theory identifies four different patterns of technology transfer for explicit and tacit technology.

- Secondly, you will learn about *absorptive capacity* (Section 4.2). This concept explains the capability of a firm or country to recognize and assimilate new technology.

- Thirdly, we will introduce a *social network perspective* on technology transfer, focusing on the social relationships between individuals as an influencing factor on the search and transfer of technologies (Section 4.3).

We conclude this chapter illustrating all three theories with an in-depth case study at Hyundai Motors (Section 4.4) and deriving implications for practice (Section 4.5).

4.1 The Theory of Technology Conversion

In Chapter 2 we learned the difference between explicit and tacit technology. The differentiation between these two forms of technology plays a significant role in the process of transferring technology. In 10 years of research Nonaka (1994), on whose work this section is based, developed a theory relating the type of technology to the process of transfer.

Four different patterns of interaction between explicit and tacit technology are identified. Each pattern formulates ways in which existing technology can be transferred and converted into new technology.

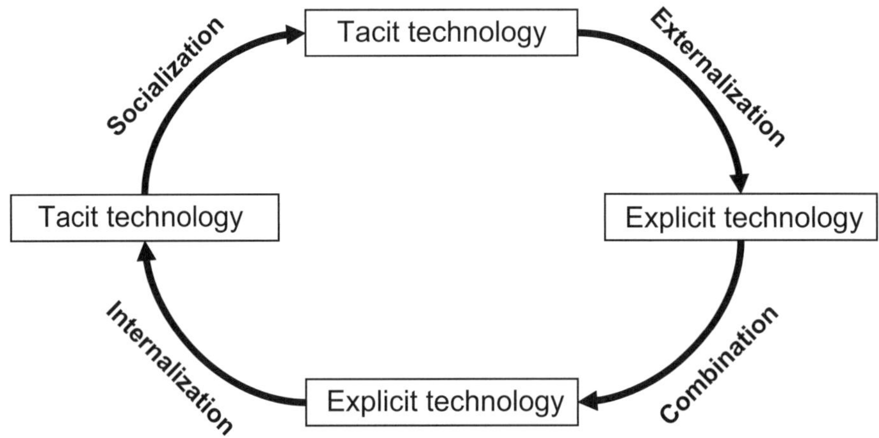

Figure 4.1   Modes of technology conversion

The four patterns of transfer (see Figure 4.1) are:

- from tacit technology to tacit technology (socialization);

- from tacit technology to explicit technology (externalization);

- from explicit technology to explicit technology (combination); and

- from explicit technology to tacit technology (internalization).

The first mode refers to the transfer of tacit technology through interaction between individuals. This is a process called *socialization* in which individuals can acquire tacit technology without language. For example, apprentices work with their mentor learning craftsmanship through observation, imitation and practice. In a business setting, on-the-job training uses the same principle. The key to acquiring tacit knowledge is experience. Without shared experience, it is extremely difficult for people to share each others' thinking processes. The mere transfer of information will often make little sense if it is abstracted from embedded emotions and contexts that are associated with the shared experience.

The second mode of transfer relates to the conversion of tacit to explicit technology. This pattern is called *externalization*. It takes place when an individual is able to articulate the foundations of their tacit technology. This transformation may be achieved through active reflection, dialogue, or thinking. It is critical for a firm or region because only externalized knowledge may be accessible by a greater variety of individuals. An example is the documentation of best sales practices of a firm's sales department.

The third mode of technology conversion includes the transfer from explicit to explicit technology. Nonaka calls this process *combination*. Individuals exchange and combine knowledge through exchange mechanisms such as meetings and telephone conversations. The reconfiguration of existing technology through the sorting, adding and recategorizing of existing explicit technology may lead to new technology. For instance, a strategy consultant performing a competitor analysis creates new explicit technology by combining already existing explicit technology from sources such as financial models, annual reports, verbal client information and sales data. Another example is a scientist publishing a research paper that is based on a combination of different theories taken from the scientific literature.

The last pattern is the conversion of explicit to tacit technology. This process is called *internalization*. In order to act on information, individuals have to understand and internalize it. This involves creating their own tacit technology. By reading documents, they can to some extent re-experience what others previously learned. This process bears some similarity to the traditional notion of learning: thus 'learning by doing' and action is deeply related to internalization. An example is a secretary internalizing a user's manual of a software program and converting the manual's explicit technology into a tacit stock of technology.

In firms, regions or countries these four patterns are related to each other through cyclic interaction. If managed correctly, this interaction leads to a continued growth in explicit as well as tacit technology, creating a continuous cycle (see Figure 4.1). This cycle is characterized by shifts between the four modes of technology conversion. Each phase is stimulated by different triggers:

- The *socialization* mode usually starts with the building of a field of interaction among individuals. Such a field may be a team, a work group, or a department within a company.

- The *externalization* mode is triggered by successive rounds of meaningful dialogue and collective thinking. In this dialogue, metaphors can be used to enable team members to reveal hidden tacit technology that would otherwise be hard to communicate.

- The *combination* mode is facilitated by triggers such as coordination among team members and the documentation of existing knowledge. Modern information and communication technologies such as an intranet or document management systems may provide a platform for this process

- Finally, *internalization* is triggered through experimentation and 'learning by doing'. Time and space have to be provided to perform this activity.

4.2 The Theory of Absorptive Capacity

As shown in the previous section, the transfer of technology depends on the type of technology to be transferred. However, whether a company, region, or country may acquire technology is also a function of its capacity to absorb technology. For example, a firm producing plastic toys in Thailand will have fewer problems in acquiring and implementing technology for keyboard manufacturing than companies that have never worked with plastics (see for example UNCTAD 2005b). This issue is analysed using the theory of absorptive capacity developed by Cohen and Levinthal (1990). We base the explanations in this section on the contributions of these two authors.

The *absorptive capacity* of an individual, firm, industry, or region is defined as the ability to recognize the value of new technology and to assimilate it. Absorptive capacity occurs at different levels:

- The lowest level is the *individual level*. In this context, the concept refers to the learning capabilities of individual members of a firm.

- The next level is the *firm or organizational* level. A firm's absorptive capacity is not simply the sum of the absorptive capacities of its employees. It also depends on the external and internal structures of a firm, steering information and technology flows.

- Absorptive capacity may also be encountered at the *macro level*, for example the level of industries, regions, or entire countries. As with firms, the intra- and inter-industry structures are important for the capability to acquire and recognize technologies.

Absorptive capacity is mainly influenced by two factors.

First, accumulated *prior related technology* increases the ability to make sense of and to assimilate and use new technology. Psychological research suggests that accumulated prior knowledge increases both the ability to put new knowledge into memory and the ability to recall and use it. The learning of new technology is more difficult in novel domains. For example, knowledge of algebra is more useful in the learning and use of calculus than it in learning and use of world history. It is the prior possession of relevant technology and skill that gives rise to creativity, permitting the sorts of associations and linkages that may never have been considered before. Prior related technology always needs to be assessed in relation to task difficulty. Hence, relevant prior technology may be basic skills and general technology in the case of developing countries, but include the most recent scientific and technological knowledge in the case of industrially advanced countries (Kim 1998: 507).

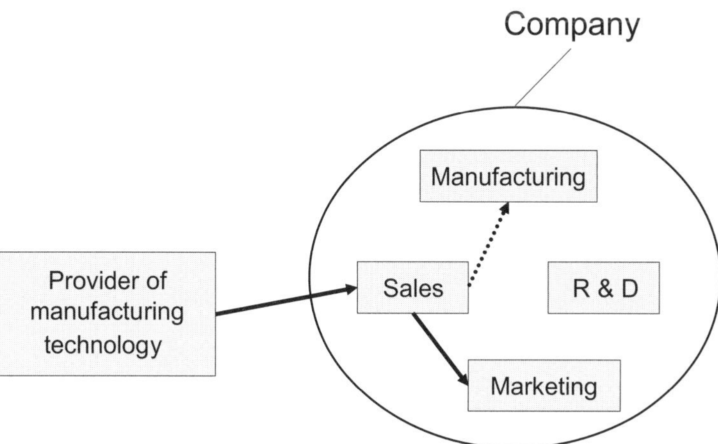

Figure 4.2   **Internal and external communication channels**

Secondly, *external and internal structures* affect the absorptive capacity of a firm, industry, region, or country. This includes the existing communication channels across firms/industries or between subunits. A firm's or region's absorptive capacity depends on the individuals that stand at the interface of their firm or region with the external environment or at the interface between departments within the firm. For example, a sales representative may encounter a new manufacturing technology at a conference they are attending. However, if no sufficient channels exist between the sales representative and the manufacturing department, this technology may never be absorbed in the production process of the firm (see Figure 4.2).

Table 4.1   Factors influencing absorptive capacity

		Communication structures	
		Open	**Closed**
Prior technology base	**High**	1. High and further increasing absorptive capacity	2. Decreasing absorptive capacity
	Low	3. Increasing absorptive capacity	4. Low and further decreasing absorptive capacity

As shown in Table 4.1, prior related technology and communication structures indicate the level and development of absorptive capacity (for a similar conceptualization see Kim 1998: 508). When prior technology is high and communication structures are open (1), absorptive capacity is high and further increasing. When the prior technology base is low and communication structures are closed (4), absorptive capacity is low and further decreasing. Firms or industries with a high prior technology base and closed communication structures (2) will gradually lose their absorptive capacity, moving down to (4). This is because their prior technology base will become obsolete as task-related technology is progressing. In contrast, organizations with low prior technology and open communication structures (3) will be able to acquire additional absorptive capacity, moving progressively to (1). This is because technology acquired through open channels will increase the existing absorptive capacity.

4.3 Social Network Theory

The theory of absorptive capacity explains the capability of a firm or nation to absorb technology as a function of its prior technology base as well as its communication channels. Social network theory makes a contribution in analysing these communication channels and explaining their importance for the transfer of explicit and tacit technology.

A *social network* is a set of actors being connected by a set of relationships (Mitchell 1969: 2, Borgatti and Foster 2003: 992, Ricken 2005, Ricken and Seidl 2010). Actors can be individuals, teams, departments, or entire companies. The content of relationships can be communication, friendship, advice or the transfer of resources, knowledge and technology.

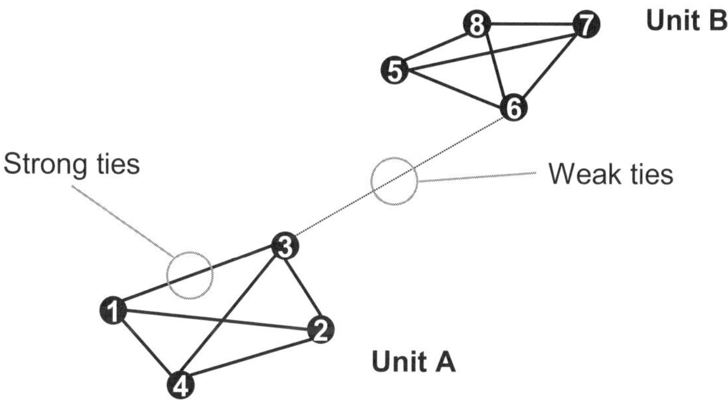

Figure 4.3   Strong and weak ties

Relationships in social networks differ by the strengths or their ties (see Figure 4.3 and for the following Granovetter 1973, 1982, Hansen 1999):

- *Strong ties* connect actors that interact frequently. Maintaining strong ties requires a significant amount of time and cost, such as frequent visits and meetings with other people. Strong ties are often redundant. For example, a group of engineers working frequently with another group of engineers is, over time, likely to be introduced to working relationships held by the other group of engineers. This results in a circle of engineers who all know one another. In the Figure 4.3 actor 1 holds a relation with actor 3 but could also reach him via actor 2 or 4: the relation is therefore to some degree redundant.

- *Weak ties* are relationships characterized by low frequency of interaction. Weak ties are less cost-intensive than strong ties. In addition, they are usually not redundant as they often reach out to distant regions of a network. For example, in Figure 4.3, actor 3 has only one possibility to reach actor 6 and his unit, meaning the relationship is not redundant.

Strong and weak ties influence the search for and the transfer of new technology (see Hansen 1999: 84–89):

- Individuals and units with *weak ties*, that is infrequent relationships, are likely to have a more advantageous search position for new technology than units having strong ties. As they are less costly than strong ties, a unit may afford to maintain a larger quantity of these ties. Weak ties can therefore offer access to distant regions of a network, providing a unit with new, non-redundant technologies.

- On the other hand, *strong ties* facilitate the transfer of tacit technology. As tacit technology is hard to verbalize, it is usually transferred through socialization (see Section 4.1), requiring frequent interaction. In addition, the knowledge source is likely to be more willing to transfer the knowledge if it is already in close interaction with the other person.

In summary, as long as firms and nations are still in the search for new technologies, they can use their time and resources to establish a large number of weak ties to different institutions and organizations. However, once a technology of interest has been identified, they need to focus their efforts and establish close interaction (strong ties) to the transferring party in order to acquire the technology of interest.

4.4 Case Study: Technology Transfer to Korea's Automotive Industry

We will now illustrate these theoretical concepts with a case study on Hyundai, a Korean automotive manufacturer (the following case is partly based on Kim 1997, 1998). We will describe how Hyundai acquired assembling and manufacturing technology (see Section 2.5), facilitating an increase in absorptive capacity (see Section 4.2).

4.4.1 TRANSFER OF ASSEMBLING TECHNOLOGY

In 1968 Hyundai wanted to acquire assembling technology and entered into an assembler agreement with Ford, assembling Ford compact cars. As part of this agreement, Ford transferred *explicit technology* to Hyundai, such as blueprints, technical specifications and production manuals. However, assembling also made the transfer of *tacit technology necessary*, such as engineering skills and know-how. The required frequent interaction between the transferring parties (see Section 4.3) was established through the following measures (Kim 1998: 510):

- First, Hyundai engineers were trained at Ford sites, initiating a process of *socialization* (see Figure 4.1).

- Secondly, Hyundai speed up the process of *internalization* of explicit technology (see Figure 4.1) because its production members applied the trigger 'learning by doing'. While plant construction was underway, production teams rehearsed production operations by disassembling and reassembling two passenger cars, a bus and a truck again and again, thus *internalizing* transferred explicit technology (production manuals) into tacit technology (see Figure 4.1).

Through these open communication structures to Ford (see Section 4.2) and the frequent interaction via training (see Section 4.3), Hyundai was able to acquire assembly technology. By opening up communication channels, the company raises its absorptive capacity (see Table 4.1).

4.4.2 TRANSFER OF MANUFACTURING TECHNOLOGY

In 1973 Hyundai wanted to develop a 'Korean' subcompact car that could be exported in substantial volume and simultaneously increase local market share. This required the acquisition of manufacturing technology (Kim 1998: 511–514).

- As a first step, Hyundai initiated an *internalization process* (see Figure 4.1). It had a project team master literature related to the various aspects of auto design and manufacture. The project team accumulating new tacit technology converted from explicit literature knowledge to enhance its prior technology base.

- Afterwards, Hyundai entered into a *licensing agreement* with Italdesign for body styling and design. In order to absorb design technology, Hyundai formed a team of five design engineers and had them study auto styling literature (internalization). Later, Hyundai sent the team to Italy to participate closely with Italdesign engineers in the design process (socialization), establishing frequent interactions and allowing the transfer of tacit technology (see Section 4.3). For one-and-a-half years, the engineers lived together in an apartment near Italdesign, kept a record of what they were learning during the day (externalization), and had group reviews every evening (combination). These intensive interactions resulted in a very rapid spiral process of technology conversion within the team, from socialization to externalization, combination and internalization (see Figure 4.1). To establish strong ties to the rest of the company (see Section 4.3), the team engineers later became the core of the style design department at Hyundai.

In summary, at the outset of phase 2, prior technology in relation to task difficulty (manufacturing) was low (see Table 4.1). Through the measures described above, the firm opened communication structures. This allowed the firm to increase its prior technology base and absorptive capacity (see Table 4.1).

As a result Hyundai developed its first indigenous model, Pony, with a 90 per cent domestic content in 1975, making Korea the second nation in Asia to have its own domestic automobile. Hyundai quickly improved the car's quality over the years. It exported 62,592 cars to Europe, the Middle East, and

Asia, accounting for 67 per cent of Korea's total auto exports in 1976–1980. Pony accounted for 98 per cent of Hyundai's exports during those periods. The company's local market share in passenger cars also increased, from 19.2 per cent in 1970 to 73.9 per cent in 1979 (Kim 1998: 512).

4.5 Key Implications for Practice

The Hyundai case study illustrates the three theories described in Sections 4.1 to 4.3. We want to close this chapter by summarizing the practical implications of the three theories for technology transfer:

- First, managers need to gain an understanding of the type of technology they are planning to transfer. As the Hyundai case shows, there is a difference in transferring explicit or tacit technology and each form of technology requires different triggers. At Hyundai, the explicit technology on engines and transmissions stored in documentations was acquired by simply studying and analysing these documents. To internalize this technology, the engineers needed to apply and try out this knowledge in practice by experimenting and learning by doing.

- Secondly, managers need to perform an analysis of the absorptive capacity of their region or country regarding a certain technology. As the Hyundai case shows, an organization is very unlikely to directly jump from capabilities of assembling automobiles to innovative capabilities in designing automobiles. Rather, companies and countries are only able to gradually increase their absorptive capacity, taking steps on the ladder of technological capabilities one after the other. Managers therefore need to make a judgement as to whether the prior technology base of their country is sufficient for the targeted technology or not.

- Finally, managers have to establish the necessary interactions for the transfer of technology. In particular, the transfer of tacit technology such as know-how and skills requires very frequent interaction between the two transferring parties. Managers have to establish platforms of time and space for such interactions.

5

Foreign Direct Investment

In Chapters 2 to 4 we developed a profound understanding of technology and technology transfer. We now want to develop a meaningful understanding of foreign direct investment (FDI): FDI is generally acknowledged as one of the most important mechanisms for technology transfer. Many regions and countries strive to attract FDI simply to absorb and retain new technologies:

- We start this chapter by illustrating *historical and recent developments in FDI* (Section 5.1).

- We deliver a *definition of FDI* and explain its components (Section 5.2).

- Subsequently, we give a description of the various *forms of FDI* (Section 5.3).

- In addition, we describe *the various determents influencing the* FDI decision of a multinational corporation (Section 5.4).

- Finally, we lay out the *contribution of FDI* to local economies (Section 5.5).

5.1 Historical and Recent Developments in FDI

Foreign direct investment is not a new phenomenon. International organizations have been engaged in trading activities from as far back as 2500BC. Early resemblance to the modern foreign direct investor appeared in the 1600s and 1700s, when companies such as the British East India Company and the Dutch East India Company entered parts of Asia, the Indies and America (Grant 1991: 284, Jones and Wren 2006: 12).

However, the appearance of modern FDI incorporating control over foreign production units did not occur until the beginning of the nineteenth century, and it was not until the latter part of the nineteenth century that larger-scale FDIs were to occur (Jones and Wren 2006: 12). These were partly driven by new improvements in transportation and communication through railways and telegraphs (Jones and Wren 2006: 12).

After the Second World War a new wave of FDI emerged, arising mainly from the US to Europe due increased political stability and a liberalized attitude towards investments.

However, it was not until the latter part of the twentieth century that world FDI flows began to increase substantially. Over the last 30 years, FDI has grown strongly at 13.9 per cent p.a. (see Figure 5.1).[1] Factors behind this include global GDP growth, low interest rates, the increasing profits of MNCs, greater availability of capital, trade liberalizations and intensified policy efforts by governments to attract FDI. In 2007, FDI inflows in all countries reached a historic peak of ~2 trillion US$, thereafter dropping to ~1.7 trillion US$ in 2008 as a result of the global financial crises. With ~1 trillion US$ or 56 per cent of the total (2008), the major share of FDI inflows still goes to developed countries. Nevertheless, the contribution of transition and developing economies is continuously increasing and has reached 44 per cent in 2008, compared to only 17 per cent in 1990.

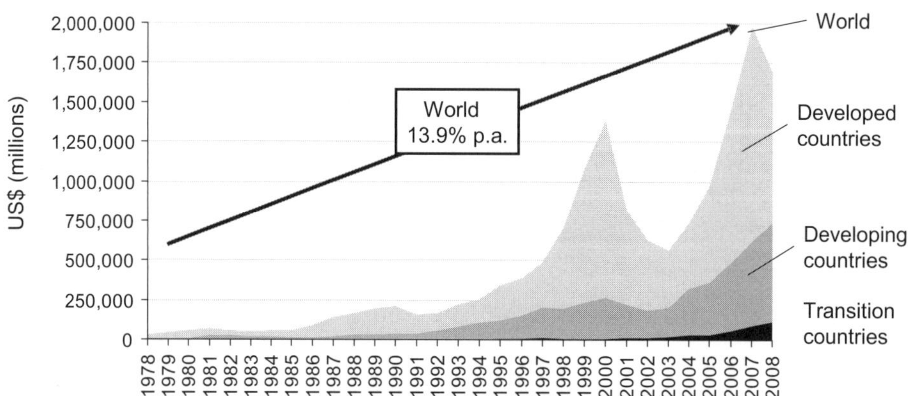

Figure 5.1   Inward FDI in US$ at current prices and exchange rates

Data source: UNCTAD.

---

1   Compound annual growth rate (CAGR) 1978–2008, based on US$ at current prices and current exchange rates.

In addition, as Figure 5.2 shows, FDI flows are very volatile and depend largely on global economic activity. For example, FDI flows dropped in 1991, 2001 and 2008, when global economic growth declined and several countries were in an economic recession. There are various reasons for this dependency of FDI on global economic activity (UNCTAD 2009: 5–8):

- *Reduced access to finance*: especially in the financial crisis of 2008, multinational corporations have experienced a limited capacity to invest, due to tighter credit conditions and lower corporate profits.

- *Negative market outlook*: recessions impact companies' forecasts on future domestic and foreign demand, decreasing the requirement of expansion in production capacity.

- *Risk aversion*: companies' investment plans are also scaled back due to a high level of perceived risks and uncertainties in times of economic tensions.

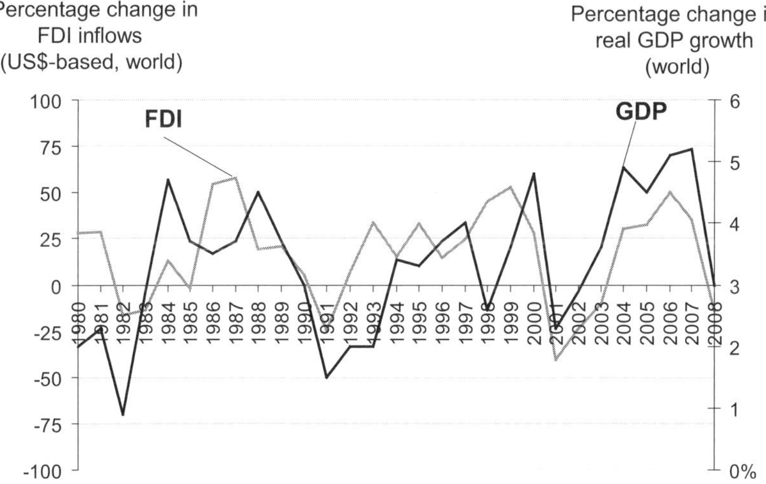

Figure 5.2   Percentage change in FDI inflows and real GDP
Data sources: IMF, UNCTAD.

5.2 Definition of FDI

'Foreign direct investment is defined as investment by a resident entity in one economy with the objective of obtaining a *lasting interest* in an enterprise resident in another economy' (OECD 2009: 88). The lasting interest means the existence of a long-term relationship between the direct investor and the enterprise. A significant degree of influence by the direct investor on the management of the direct investment enterprise is also required. The basic criterion for *influence* used by international organizations such as OECD is the ownership of at least 10 per cent of the voting power (OECD 2009: 88). Control by the foreign investor is not required but lower shares than 10 per cent count as *portfolio investments*.

FDI is usually undertaken by a multinational corporation (MNC). MNCs are companies that have established operating units via FDI in at least two countries (Jindra 2006: 6). Via FDI, MNCs establish a so called *foreign or local subsidiary*.[2] The MNC or investor owning this foreign subsidiary is usually called the *parent company* (Jindra 2006: 6). An investment involves both the initial transaction between the parent and the subsidiary and all subsequent transactions.

FDI has three components (UNCTAD 2009: 243):

- *Equity capital* is the foreign direct investor's purchase of shares of an enterprise in a foreign country.

- *Reinvested earnings* comprise the direct investor's share (in proportion to direct equity participation) of earnings not distributed as dividends by affiliates, or earnings not remitted to the direct investor. Such retained profits by affiliates are reinvested.

- *Intra-company* loans or intra-company debt transactions refer to short- or long-term borrowing and lending of funds between parent enterprises and affiliate enterprises.

5.3 Types and Forms of FDI

There are basically two types of FDI:

---

2   In this book we use the term foreign or local subsidiary to embrace all companies having at least 10 per cent foreign capital (see also Jindra 2006: 7).

1.   *Greenfield investments* take place in new facilities or the expansion of existing facilities. An example is a beverage company establishing a new production unit in Nigeria in order to produce and sell beverages for the Nigerian market. Greenfield investments are the primary target of a host nation's promotional efforts because they create new production capacity, jobs and transfer technology and know-how. However, in the case of establishing an entirely new subsidiary, greenfield investments also involve a high risk for the investor who will have to build up an entirely new unit and faces the risk of not being able to build relationships with customers or suppliers or not being able to recruit the necessary personal (Mellahi et al. 2005: 196).

2.   *Mergers and acquisitions* take place through a transfer of existing assets from local firms to foreign companies. Cross-border *mergers* occur when the assets and operation of firms from different countries are combined to establish a new legal entity. Cross-border *acquisitions* take place when the control of assets and operations is transferred from a local to a foreign company, with the local company becoming an affiliate of the foreign company. Typical risks of performing mergers and acquisitions are the integration of the acquired company causing problems due to cultural, structural, technological, or procedural obstacles (Mellahi 2005: 198). A famous example of the materialization of such risks was the acquisition of Chrysler by Daimler Benz in 1998, eventually resulting in the divestment of the acquired unit in 2007.

FDI may also take different forms (see also Cohen 2004: 127):

- A *wholly owned subsidiary* is a company controlled by another company or corporation. Subsidiaries are separate, distinct legal entities for the purposes of taxation and regulation. They are distinct from divisions, which are entities fully integrated within the main company, and not legally or otherwise distinct from it.

- A *joint venture* (JV) is a strategic alliance between two or more parties to undertake economic activity together. The parties agree to create a new entity together by both contributing equity, and then share the revenues, expenses and control of the enterprise. The venture can be for one specific project only, or a continuing business relationship. Other than wholly owned subsidiaries the partners in a joint venture are and remain independent units with partly common but partly conflicting objectives (Navaretti and Bigano 1998: 44).

- *Minority holdings* are investments by a foreign direct investor that exceed that threshold of 10 per cent (see Section 5.2) and allow some influence on the local subsidiary. Nevertheless, the foreign investor does not exceed full or joined control over the company. An example would be a MNC acquiring a 20 per cent 'seed investment' in a local telecommunications company with the remaining 80 per cent owned by a local investor. In this case, the foreign investor does not exceed full control but may still influence the business, for example by nominating one or two persons for the board of directors. MNCs might use such minority investments to gain a first foothold into a country while later expanding their participation. Also, some countries such as China exceed restrictions on majority holdings.

The different forms of FDI may be used to realize horizontal as well as vertical investments (see Navaretti and Bigano 1998: 42):

- *Horizontal investments* refer to the reproduction abroad of company processes which are already carried out home. Examples are Volkswagen's automotive operations in Brazil and Mexico producing for the Latin American market.

- *Vertical investments* relate to the setting up of business processes which are not carried out at home. For instance Cemex, a Mexican building materials firm, has offshored parts of its R&D activities to Switzerland due to better access to qualified labour and closeness to technical universities. Another example are Swiss banks, such as Credit Suisse, having offshored parts of their human resource activities to Eastern European countries such as Poland. Similarly, some large consulting companies such as McKinsey have allocated a share of their back-office processes to India.

5.4 Determinants of FDI

In order for nations to attract technology via FDI, they have to understand the determinants of an investment decision of a foreign direct investor. In this chapter we therefore take the perspective of a foreign direct investor, mostly a MNC, and analyse the motives to perform a foreign direct investment. We thereby want to focus on two questions:

- Why do MNCs internationalize and perform FDI in general? (Section 5.4.1).

- What determinants influence the management of a MNC to prefer a certain FDI opportunity over another? (Section 5.4.2).

5.4.1 WHY DO FIRMS PERFORM FDI IN GENERAL?

Local companies usually have a better understanding of the local market, the local customers and other country-specific conditions. MNCs are able to compete successfully with local competitors because they are often in possession of advanced technologies, management, marketing know-how and economies of scale. But what motivates multinational corporations to internationalize and perform FDI in the first place? There is a variety of reasons (see Grant 1991: 284):

- *Acquisition of resources*: a company may decide to invest in a foreign country in order to exploit resources which it does not possess in its home country. For example, in the nineteenth century industrial companies in Europe and North America moved overseas to exploit raw materials such as oil, bauxite, rubber, or iron ore.

- *Exploitation of country-specific factors*: companies may also decide to perform FDI in order to take advantage of country-specific production factors. For instance, global toy and textile manufacturers have established production units in China because of lower labour costs. Such expansions can also be directed towards the acquisition of knowledge or technological skills: European and Japanese IT companies have established research units in California's Silicon Valley because they want to take advantage of the available qualified labour and the benefits of industry clusters (see also Section 12.2.1).

For similar reasons, US textile manufacturers have established design studios in Italy.

- *Realize profit and growth opportunities*: a firm may decide to invest abroad because a foreign country can offer better opportunities for growth or profit than the domestic market. For example, a weak domestic economy may motivate a MNC to seek FDI. Also, a MNC may go abroad because a domestic market may have reached saturation for a certain product.

- *Risk diversification*: the internationalization of operations permits the spreading of country-specific risks. The more risky the domestic economy is perceived to be, the more attractive FDI becomes compared to domestic investments. For example, in the global economic crises of 2008 and 2009, many MNCs were able to stabilize their earnings because of their global presence. While volumes and prices in Europe and the United States were often negatively affected by the global recession, many emerging markets, such as India, Brazil, or China performed reasonably well. Another phenomenon in this context are emerging market companies such as the Indian steel producer Mittal Steel consequently seeking FDIs in the United States and Europe in order to reduce their emerging market risk.

- *Overcoming trade barriers*: FDI is an alternative to exporting. The greater the cost of exporting, either from transportation or tariffs, the more attractive it becomes to establish a foreign production unit. For instance, the imposition of tariffs and quota restrictions on Japanese automobile exports to the United States was a principal factor in establishing Japanese plants in the United States.

5.4.2 WHAT INFLUENCES A MNC'S DECISION FOR A PARTICULAR FDI OPTION?

In the previous Section we have described general reasons why a company will internationalize and perform FDI. For managers of government bodies and investment promotion agencies it is equally important to understand why MNCs may decide to follow one FDI opportunity over another. Broadly speaking, three categories of determents can be identified:

1.   A MNC's strategy.

2.   The expected financial return of an investment opportunity.

3.   Other factors.

5.4.2.1 MNC strategy

A MNC will become involved in a particular FDI opportunity in case the opportunity matches its global strategy. The geographic, product or cost strategies of a MNC may each contain elements that can influence a MNC's approach towards a particular FDI opportunity. For example, Orascom Cement, an emerging markets cement manufacturer, pursued the geographic strategy of investment in high-risk and high-return counties. Therefore, the company pursued large investment projects in countries such as Iraq and Pakistan but did not focus on opportunities in industrialized countries such as Germany or the United States. For government and investment promotion agencies, understanding such strategies is a critical precondition to developing a plan for attracting FDI.

5.4.2.2 Expected financial return

An investor will also become involved in a FDI project when the expected financial return (measured by the net present value) of such a project is both positive and greater than those of alternative options for investment (Rivoli and Salorio 1996: 336, see also Section 10.2.5). The following factors influence the return of an investment and are usually analysed by the management of a MNC (see also Porter 1990: 71, Ikiara 2003: 21, Jones and Wren 2006: 46–48):

- The *macroeconomic conditions* of a country exceed an influence on the overall economic situation of a country. Conditions include the size and growth of the economy, interest and exchange rates, political stability, fiscal policy and corruption.

- The *attractiveness of the local market* is relevant for investments targeting the local economy. Market attractiveness is influenced by factors such as the size and growth of the market, the prices for products and services, entry barriers or the number and behaviour of competitors. For an electricity company, for instance, a large and fast-growing market such as India, China or Indonesia can become attractive simply because of the expected absolute increase in domestic demand for electricity.

- *Factor conditions* such as human resources (cost, quality and availability of labour), physical resources (land, water, minerals, power, etc.), infrastructure (roads, railway, seaports, airports, etc.), the geographical proximity to markets (logistic costs) and the availability, costs and quality of suppliers are other significant issues. Investors will analyse these factors because they exert an influence on both the quality of a product as well as its production costs. For example, Singapore is an attractive destination for export-oriented FDI due to its excellent port infrastructure and logistics position for seaborne trade. Similarly, many manufacturers have outsourced their production units to Asia due to the cheap labour in these countries.

- *Investment and trade policies* such as attitude towards FDI, bureaucracy, transparency and guarantee of property rights influence the risk a MNC takes with its investment decision. Even if a domestic market is attractive, MNCs may chose not to perform any direct investments because of unfavourable policy conditions, for example the threat of expropriation, unfavourable taxes, low security standards, or theft of technology. Also, a country's trade policy impacts the MNC's choice between exporting to a market and producing locally. For example, tariff barriers on the import of automobiles might encourage international manufacturers to set up local production subsidiaries and could increase FDI inflows. However, in the long run, such protectionism might negatively affect the economic performance.

5.4.2.3 Other factors

Besides the investor's strategy and the expected financial return of an option, other factors may influence the investment decision. Language ties and cultural similarities are notable determinants on a MNCs decision. For example, many Spanish multinationals such as Telefónica, Iberia, BBVA, or Repsol YPF have predominately realized FDI projects in the culturally related countries of Latin America. Other influencing factors are personal or emotional ties of an investor to a certain country. A typical example is the Orascom Group, a large Egyptian MNC currently performing a billion dollar investment in a hotel resort in

Andermatt, Switzerland. One potential influencing factor on this FDI decision could be that the owning brothers have ties to German-speaking countries as they went to a German school in Cairo and have studied in Switzerland and Germany.

5.5 Potential Contributions of FDI

FDI is considered an important pillar of economic development. There are multiple avenues through which FDI can contribute to the economy (Moran 2005: 300, Morais da Costa et al. 2006: 214, Akrami 2008: 80):

- The foreign investor can *provide a new product* that the local companies are not capable of supplying or the investor might provide an existing product of improved quality or at a lower price. An example is an international telecommunications firm setting up a mobile communications network in an African country which previously did not dispose of mobile telephony.

- FDI can *compensate the low level of domestic savings*, contributing to the accumulation of capital and compensating domestic capital deficits. For example, a Chinese mining company setting up mining activities in Afghanistan provides capital to this country which might not be available locally.

- Foreign direct investors can stimulate the economy by *creating an additional demand from* their local subsidies for host country suppliers and labour. For instance, a foreign investor constructing a coal-fired power plant in Indonesia will generate additional demand by sourcing coal from Indonesian mines and using domestic labour.

- FDI might allow the host country to penetrate international markets, *earning foreign exchange* and/or allow competitive substitution of import. For example, an international investor setting up flower plantations in Ecuador and exporting flowers to Europe will generate an additional foreign currency inflow for the country.

- FDI has the potential to *supply knowledge and technology*, allowing the host country to engage in existing activities more efficiently or to undertake entirely new activities. An example is a textile manufacturer setting up production facilities in Vietnam, transfering knowledge on quality control, production and process techniques, as well as management know-how.

6

FDI and Technology Transfer

As mentioned earlier, one of the major expected benefits of FDI is the transfer of technology. Although trade and licensing agreements are other channels through which technology transfer can take place, FDI is generally acknowledged as one of the most important mechanisms. Empirical studies have found positive and significant knowledge spillovers stemming from inward FDI (Javorcik 2004, Bitzer and Kerekes 2008). One reason is that FDI is usually brought by MNCs which exhibit a high ratio of R&D (Saggi 2000: 17).

Table 6.1 **FDI and technology transfer. To what extent does FDI bring new technology into your country? 1, not at all; 7, FDI is a key source of new technology**

Top 5 ranked countries			Bottom 5 ranked countries		
Rank	**Country**	**Score**	**Rank**	**Country**	**Score**
1	Ireland	6.3	129	Nepal	3.6
2	Singapore	6.2	130	Chad	3.5
3	Luxembourg	5.9	131	Bolivia	3.3
4	Slovak Republic	5.9	132	Zimbabwe	3.1
5	Qatar	5.7	133	Algeria	3.0

Data source: WEF (2009: 443).

However, the degree to which FDI results in an actual transfer of technology seems to vary. Table 6.1 indicates that for countries such as Ireland, Singapore, Luxembourg, the Slovak Republic or Qatar FDI seems to be an important source of new technologies. However, countries such as Algeria, Zimbabwe, Bolivia, Chad, or Nepal do not seem to experience significant technology transfer via FDI (see WEF 2009: 443). A critical driver for these considerable differences is the functioning of the various forms of technology transfer. In general, three forms of technology transfer via FDI exist (see Figure 6.1) (Blalock and Gertler 2005: 77, Jindra 2006: 10):

1.   *Direct, internal transfer* from the MNC to its local subsidiary (see Section 6.1);

2.   *Horizontal, external spillovers* to local competitors (see Section 6.2); and

3.   *Vertical, external spillovers* to backwardly linked suppliers or forwardly linked customers (see Section 6.3).

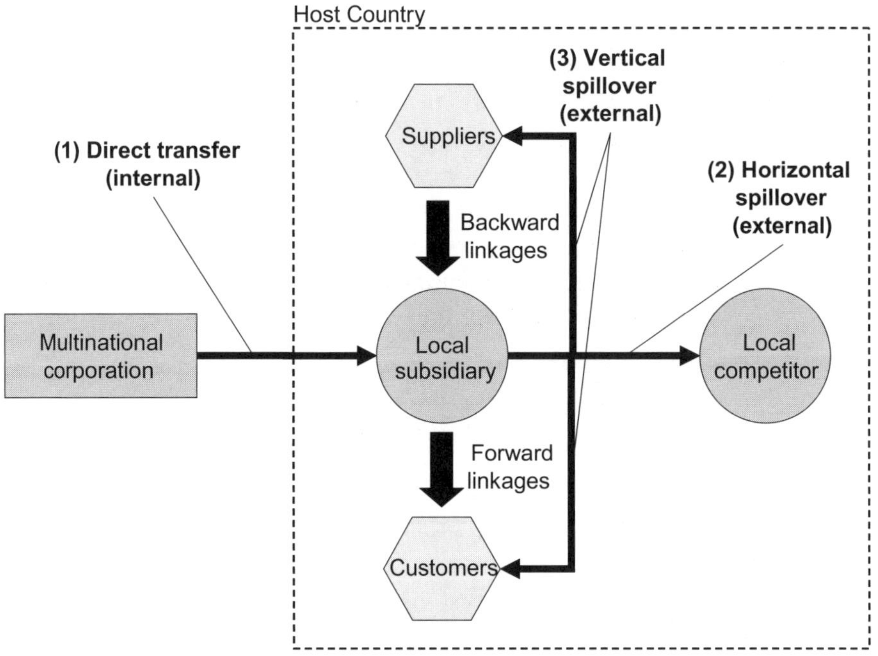

Figure 6.1   Three forms of technology transfer in FDI

6.1 Internal Transfer from the MNC to its Local Subsidiary

Direct technology transfer refers to the internal transfer of technology from a parent MNC to a local subsidiary in a host country (see Figure 6.1).

The benefits of such as transfer can be fully internalized in the form of efficiency enhancement by the foreign subsidiary.

There are a variety of factors which influence a MNC's decision as to whether it wants to transfer technology to its subsidiary or not:

- The *global strategy* of the MNC and the defined role of the local subsidiary. Factors such as the desired autonomy of a subsidiary or its specific objective exert an influence on the required transfer of technology from the parent company.

- The *expected benefits* of a transfer of technology to the subsidiary. A MNC may transfer product technology in order to enhance the quality of the products sold in the local market. Or MNCs may transfer process technologies, in order to produce more cost-efficiently in the host country.

- The *absorptive capacity* of the local subsidiary in recognising the values of the new technology and using it (see Section 4.2). For example, a MNC will not transfer innovative capabilities for product development if its foreign subsidiary is merely in possession of assembly capabilities (see Section 2.5).

- The *risks of technology transfer*, mainly lying in the possibility of technology theft or unwanted spillovers to competitors or other organizations.

- *Policy factors* such as local financial and fiscal incentives for technology transfer.

Mechanisms through which a transfer from a parent MNC to a foreign subsidiary may take place are:

- Training of local personal by the parent company (on and off the job).

- Provision of information and documents, for example on patents, product documentation, process descriptions, etc.

- corporate rules and regulations such as accounting standards, codes of conduct, etc.

- exchange of personnel (both from the parent company to the local subsidiary and vice versa).

- provision of machinery and equipment, in the form of laboratories, production facilities, etc.

- dissemination of management tools, such as total quality management, marketing tools, forecasting and planning methods and others.

- internal consulting services from the parent company, via strategy projects, manufacturing process optimization projects, marketing reviews, etc.

6.2 External Transfer 1: Horizontal Spillover to Competitors

Technology transfer does not only occur internally between a parent MNC and its foreign subsidiary but also externally from the foreign subsidiary to other foreign firms of the host economy. Such external transfers are called spillovers. They can materialize either *horizontally* to competitors or *vertically* to suppliers or buyers. MNCs have an incentive to prevent horizontal spillovers to local competitors as this might weaken the MNCs competitive position. At the same time, MNCs often benefit from technology transfer to their suppliers. Therefore, spillovers from FDI are more likely to be vertically than horizontally in nature (Javorcik 2004: 606).

For horizontal spillovers to local competitors, two main mechanisms exist (see for the following Blomström and Kokko 1998: 287–298, Saggi 2000: 18–26, Jindra 2006: 12–16):

- *Demonstration effects*. Local competitors may adopt technologies introduced by multinational firms through imitation or reverse engineering. The demonstration effect builds on the assumption that it is often too costly for local firms to acquire the necessary information on new technologies if they do not yet exist in the local market. Geographical proximity may indeed be crucial for developing countries that are not well integrated into the world economy and may have few alternatives of absorbing technologies.

- *Labour turnover.* Workers trained or previously employed by MNCs may transfer information, knowledge and skills to local competitors by switching employers. They may also contribute to technology diffusion by starting their own firms. For example, in Tawian in the mid-1980s almost 50 per cent of all engineers and approximately 63 per cent of all skilled workers left multinationals to join local Taiwanese firms. In more developed countries, changes due to technology spillovers might be higher than in less developed countries because of the higher absorptive capacity of competitors of MNCs. MNCs may limit this form of technology drain by offering higher wages than local rivals. Local competition law may also affect turnover. For instance, under Bulgaria's competition law no person is permitted to join the management of a competing firm operating in the same line of business for the first three years after leaving the enterprise. Labour turnover rates may vary across industry sectors as well. For instance industries with a vast pace of technological change such as the computer industry are characterized by higher turnover rates than other industries.

- *Market entry of service firms.* Finally, multinational investments may encourage the entry of international professional service firms such accounting firms, law firms, or consultant companies. These may then further distribute their own technology and knowledge among the local enterprises of a country.

6.3 External Transfer 2: Vertical Spillover to Buyers and Suppliers

As described in the previous chapter, vertical spillover is more likely to occur than horizontal spillover. In detail, MNCs may transfer technology to firms that are (see Figure 6.1):

- suppliers via backward linkages

- buyers via forward linkages

6.3.1 BACKWARD LINKAGES TO SUPPLIERS

MNCs may facilitate the absorption of technologies via *backward* linkages to suppliers. They may help suppliers to set up production facilities, provide technical assistance to raise the quality of suppliers' products, facilitate innovation, provide training, help in management and organizational issues and assist in the purchasing of materials. For example, a Czech producer of aluminium alloy castings for the automotive industry signed a contract with a multinational customer. The staff from the multinational would visit the Czech firm's premises for two days each month over an extended period to work on improving the quality control system. As a result, the Czech firm applied these improvements to its other production lines (not serving this particular customer) and reduced the number of defective items produced (Javorcik 2004: 608, Blalock and Gertler 2005: 73).

The following *mechanisms* can lead to a backward spillover of technology (Jindra 2006: 15):

- Higher requirements regarding product quality introduced by MNCs set incentives for suppliers to *upgrade* their production technology.

- The demand for local inputs and quality requirements may induce the MNC to *directly transfer technology* to the supplier.

- *Labour turnover* can also lead to indirect technology transfer. Moving from a MNC to a supplier or customer is more likely than changing to a competitor as employment contracts often prohibit such a move.

BOX 6.1   BACKWARD LINKAGES TO SPANISH AND PORTUGUESE AUTOMOTIVE SUPPLIERS

The automotive industry is often cited as a location of backward linkages. Relationships with international automotive manufacturers are a window of opportunity for domestic suppliers. Via such relations, suppliers may gain access to new technologies allowing them to improve product quality, productivity and costs.

In a case study on Spanish and Portuguese automotive suppliers, Varum (2006) analyses the transfer of practices and quality standards such as ISO9001, total quality management, benchmarking, measuring client level satisfaction, or kaizen.

It turns out that the technology transfer between automotive manufacturers and suppliers is a complex process and its success depends on a clear strategy, the type of technology to be transferred (see Section 2.3) and the mechanisms of technology transfer.

On the one hand, automotive manufacturers mainly transferred *explicit technology* through the following mechanisms:

- Process-related specifications
- Manuals on firm specific quality systems
- Product-related specifications
- Yearly negotiated improvement plans

On the other hand, *tacit technology* was transferred via:

- Informal contact between personnel
- Joint meetings, teams, demonstrations and seminars
- Site visits and training of supplier's personnel
- Temporary on-site assistance and labour exchange

Varum (2006: 234–252).

6.3.2 FORWARD LINKAGES TO BUYERS

Forward linkages to customers can occur thanks to provision of inputs that either were previously unavailable in the country or are technologically more advanced (Javorcik 2004: 606).

Forward technology spillover may occur in the following cases (Jindra 2006: 16):

- MNCs may outsource their distribution and invest in their marketing outlets, for example automobile dealers, gas stations, restaurant chains, etc. MNCs may then provide technology and knowledge in the form of business concepts or training. Franchising is a common mechanism here.

- MNCs may provide services and assistance to the local buyers of their machinery, equipment and intermediate goods, for example, information and training on how to use a certain machinery or equipment or provide repair services. Similarly, industrial applications such as computer-based automation and information technologies might require expertise from the MNC.

PART II
The Practical Process

7

An Overview of the Practical Processes of Technology Transfer

In Part I (Chapters 2 to 6) we supplied the theoretical foundations of technology transfer and FDI. We now want to provide you with the knowledge, instruments and tools to actively manage the transfer of technology via FDI to your industry, region or country.

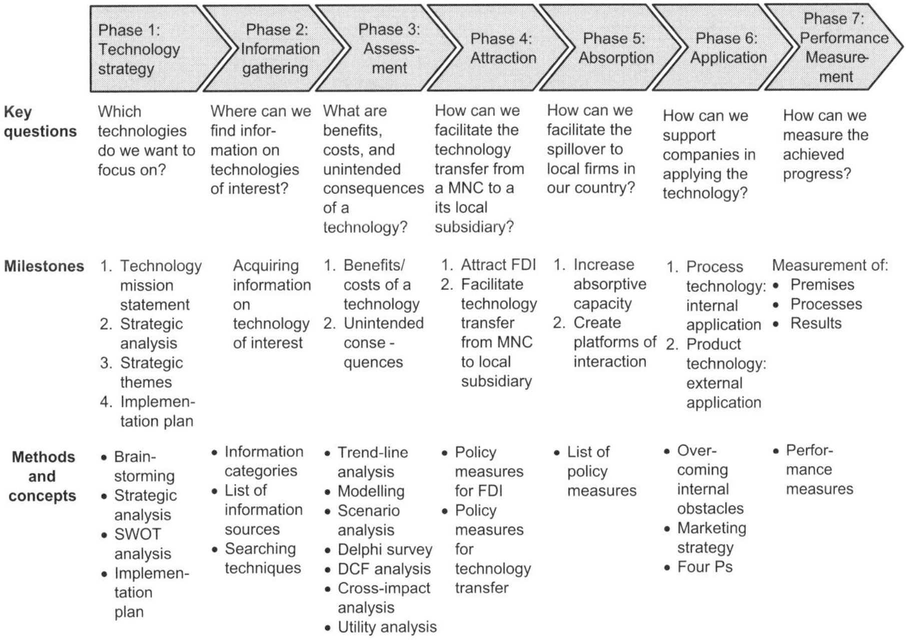

	Phase 1: Technology strategy	Phase 2: Information gathering	Phase 3: Assessment	Phase 4: Attraction	Phase 5: Absorption	Phase 6: Application	Phase 7: Performance Measurement
Key questions	Which technologies do we want to focus on?	Where can we find information on technologies of interest?	What are benefits, costs, and unintended consequences of a technology?	How can we facilitate the technology transfer from a MNC to a its local subsidiary?	How can we facilitate the spillover to local firms in our country?	How can we support companies in applying the technology?	How can we measure the achieved progress?
Milestones	1. Technology mission statement 2. Strategic analysis 3. Strategic themes 4. Implementation plan	Acquiring information on technology of interest	1. Benefits/ costs of a technology 2. Unintended conse-quences	1. Attract FDI 2. Facilitate technology transfer from MNC to local subsidiary	1. Increase absorptive capacity 2. Create platforms of interaction	1. Process technology: internal application 2. Product technology: external application	Measurement of: • Premises • Processes • Results
Methods and concepts	• Brain-storming • Strategic analysis • SWOT analysis • Implemen-tation plan	• Information categories • List of information sources • Searching techniques	• Trend-line analysis • Modelling • Scenario analysis • Delphi survey • DCF analysis • Cross-impact analysis • Utility analysis	• Policy measures for FDI • Policy measures for technology transfer	• List of policy measures	• Over-coming internal obstacles • Marketing strategy • Four Ps	• Perfor-mance measures

Figure 7.1   Managing the process of technology transfer in FDI

The process of managing technology transfer via FDI is divided into different phases (see Figure 7.1):

- *Phase 1 – development of a technology strategy* (Chapter 8). The key question managers need to answer in this phase is: which technologies do we want to attract and absorb? Answering this question involves the formulation of a technology mission statement, a strategic analysis, the formulation of strategic themes and the definition of an implementation plan. Methods such as brainstorming or strengths, weaknesses, opportunities, threats (SWOT) analysis support managers in this phase.

- *Phase 2 – gathering of information* (Chapter 9). Managers and investment promotion agencies need to gather information on the technologies of interest. We show you what information categories are relevant, from which sources information can be gathered and which searching techniques can be applied.

- *Phase 3 – technology assessment* (Chapter 10). Subsequently, managers should evaluate the technologies in question. Benefits, costs and unintended consequences of introducing certain technologies into a country need to be assessed. A variety of methods such as trend-line analysis, models, scenario building, Delphi survey, DCF analysis, cross-impact analysis or utility analysis support you in this phase.

- *Phase 4 – technology attraction* (Chapter 11). In a further step, a selected technology needs to be *attracted* to a host country. Two steps are of importance. First, managers require knowledge on how to attract a MNC and FDI to their country in general. Secondly, once an investment has taken place, managers have to develop policies stimulating the transfer of technology from the parent MNC to its established local subsidiary.

- *Phase 5 – technology absorption* (Chapter 12). Simply attracting a technology to a country or region does not yet guarantee that the technology is absorbed by local companies. For this to take place, either horizontal or vertical spillovers need to occur. We therefore illustrate a battery of policies to raise the absorptive capacity of a country and to facilitate interaction between the foreign investor and local companies.

- *Phase 6 – technology application* (Chapter 13). The application of technology in a firm, industry and country formulates the next phase of active technology management. For the internal use of process technologies, we inidcate how motivational, technological and organizational barriers can be overcome. For the external commercialization of product technologies, we show how a marketing strategy can be developed and implemented.

- *Phase 7 – performance measurement* (Chapter 13). In the last phase of the technology management process managers measure the achieved progress and learn from experience. We will provide you with the concepts and tools to measure the premises, processes and results of technology transfer.

In practice some phases might overlap or occur simultaneously. For example, you might already gather some information on technologies (phase 2) during the formulation of the technology strategy (phase 1). Also, performance measurement (phase 7) might already take place in the process of absorbing a technology (phase 4). Nevertheless, the distinction between different phases facilitates the planning of the entire project and provides conceptual guidance.

8

Developing a Technology Strategy

Strategy is the great work of the organization. In situations of life or death it is the Tao of survival or extinction. Its study cannot be neglected.

Sun Tzu, The Art of War (Wing 1998).

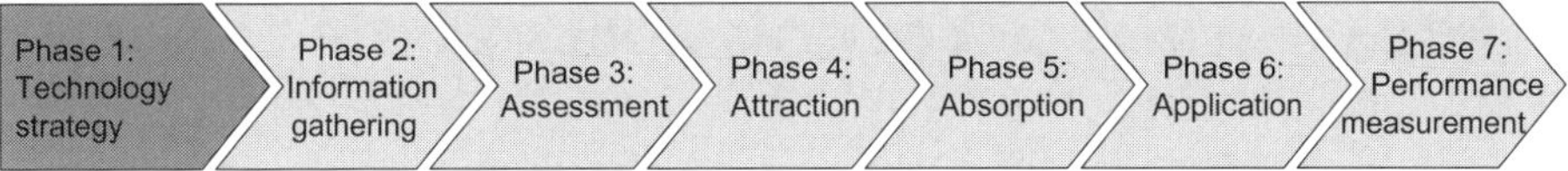

Figure 8.1   Technology strategy (phase 1)

8.1 Introduction

This chapter deals with the first phase of the process to manage technology transfer, the development of a technology strategy. A strategy 'implies the movement of an organization from its present position to a desirable but uncertain future position' (Kaplan and Norton 2001: 76).

A technology strategy is critical because it steers all the following phases and activities. Central questions managers have to answer are:

- In which technological direction should our country move in the future?

- What are the strengths and weaknesses and risks and opportunities regarding technological development?

- Which type of technologies should we focus on?

- How can we attract these technologies?

'The serious problems within the socio-economic systems of developing countries along with the absence or shortage of technological foundations and institutional requirements, mean that a laissez-faire approach will not obtain the best results: an active technological strategy is necessary' (Cohen 2004: 217).

It is noteworthy at this point that a technology strategy should not be equalized with the overall strategy of an investment promotion agency or similar organization. Rather, the technology strategy may be seen as one aspect of the overall strategy, interacting with other goals and measures of the organization. Box 8.1 gives an example of the importance of technology strategy.

BOX 8.1   KENYA'S LACK OF TECHNOLOGY STRATEGY

The UNCTAD report on the technology gap of African countries gives Kenya specific policy recommendations on technology upgrading. It diagnoses that

> *there is no institutional mechanism in Kenya for comprehensively evaluating and setting science and technology priorities for the country. There is no well developed science and technology plan, and responsibility for relevant policies is spread over a large number of uncoordinated ministries and institutions. This setting is not conducive to supporting technology upgrading in Kenya ... The first recommendation of this report is, therefore, that the Government of Kenya review and improve its strategy-making capabilities, and entrust one body with analyzing technology needs at the broad economic level and designing and implementing strategies that cut across many ministerial and departmental lines. This body should include high-level representatives from all the ministries concerned, as well as the important institutions and the private sector.*

(UNCTAD 2003c: 34, 35).

In contrast to this policy, most market- and export-oriented newly industrializing economies have had mechanisms to identify and acquire strategic technologies.

National technological priorities have been set in most industrialized countries such as France, Germany and Japan. Strategic programmes call for all parties concerned with science and technology to determine the technological course and needs for their country. A number of developing countries, including India, Korea, Thailand and several Latin American countries, are conducting similar exercises.

Adapted from UNCTAD (2003c).

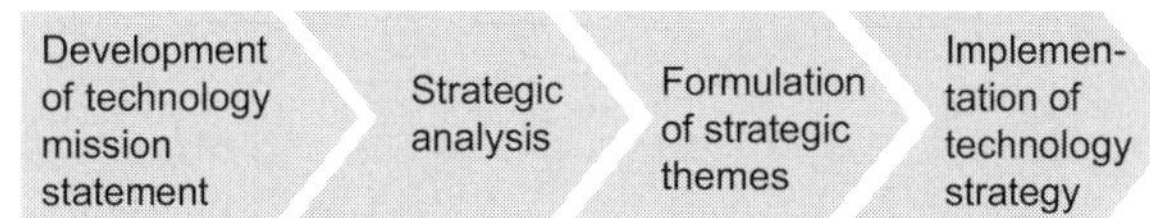

Figure 8.2   Steps of technology strategy development

Similar to the classical strategy model of the Harvard Business School (Andrews 1971) we divide the process of formulating a technology strategy into four analytical steps. Each step is related to the overall strategy of an organization (see Figure 8.2):

- The *formulation of a technology mission statement* generates an idea of the technological direction a region or country wants to take (see Section 8.2). A mission statement orientates, steers, motivates and coordinates all future activities of technology management.

- The *strategic analysis* creates transparency on the different forces influencing the economic and technological development of an industry, region or country (see Section 8.2). Before a technology strategy can be formulated, factors such as the natural resources, human resources, domestic and foreign markets and the existing technology basis should be systematically analysed. The analysis of these factors results in the identification of internal strengths and weaknesses as well as external opportunities and threats.

- The *formulation of strategic themes* refers to the planning of long-term goals for the management of technology transfer (see Section 8.4). These goals emerge from the strengths and weaknesses as well as the opportunities and threats identified in the strategic analysis.

- The *implementation of technology strategy* refers to the execution of strategic themes (see Section 8.5). This involves the actual identification, evaluation, attraction, absorption and application of new technologies.

We will now describe each phase of the process in more detail and provide you with the required tools and methods.

8.2 Development of Technology Mission Statement

The strategy process starts with the formulation of a *technology mission statement*. Such a statement needs to be aligned with the overall strategy (see Figure 8.3). It gives long-term indications of where an organization or country should go with regard to technology development and transfer. A mission statement consists of three elements:

1. A *technology mission* defines the organizations purpose and existence with regard to technology. It needs to be derived from the overall mission of an organization (see Figure 8.3).

2. The *technology core values* indicate which actions regarding technology development and transfer are viewed as appropriate and which are not. Technology core values need to be aligned with the overall core values of an organization (see Figure 8.3).

3. The *technology vision* of an organization paints a picture of the future that clarifies the direction of a company, region or country regarding its technological development. 'It launches the movement from the stability of the missions and core values to the dynamism of strategy' (Kaplan and Norton 2001: 73). The technology vision paints a picture of the future which on the one hand is close enough to grasp its possible realization. On the other hand it is far away enough to create enthusiasm among employees.

Box 8.2 gives an example of the technology mission and core values of the South African Department of Science and Technology.

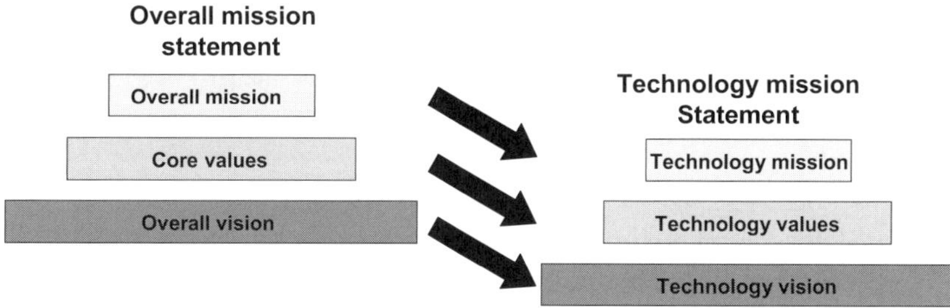

Figure 8.3   Overall and technology mission statement

BOX 8.2   TECHNOLOGY VISION, MISSION AND CORE VALUES IN SOUTH AFRICA

Mission Statement of the South African Department of Science and Technology

'Our *Vision*: To create a prosperous society that derives enduring and equitable benefits from science and technology.

Our *Mission*: To develop, coordinate and manage a national system of innovation that will bring about maximum human capital, sustainable economic growth and improved quality of life.

Core Values:

We are guided by the following values.

- *Professionalism*: An employee must strive to deliver top-class quality products and services, seek innovative ways to solve problems and enhance effectiveness and efficiency.
- *Competence*: An employee must be faithful and honest in the execution of her or his duties and must be committed through timely service towards the development and upliftment of all South Africans.
- *Integrity*: An employee must be responsible and accountable in dealing with public funds, property and other resources.
- *Transparency*: An employee must promote transparent administration and recognise the right of access to information excluding information that is specifically protected by law.

Adapted from DST (2010).

Technology mission, core values and technology vision have to be defined by the top management of an organization. A workshop provides an effective working-environment for managers to reach the necessary agreement. In such a workshop, managers can perform three steps:

- First, members of management are given time to brainstorm (5–10 minutes). Each manager writes their ideas for a technology vision on a card.

- Secondly, the cards are put on a pinboard. A discussion of the cards is initiated among workshop participants. Different views and ideas are exchanged.

- Thirdly, a collectively shared vision is formulated. First versions are written down on a pinboard and may be modified until a final version is reached. It is important that the final version is supported by all members of the top management.

The same steps can be performed to formulate values and mission.

Figure 8.4 illustrates these three steps with an example. Here, in 2003, the management of the Swiss Organisation for Facilitating Investments (SOFI), a Swiss investment promotion agency, formulated a vision for its employees:

- In a first step, the five managers participating in the workshop were given time to brainstorm. After ten minutes, the managers wrote statements such as 'continuity culture' or 'quality before quantity' on cards and put them on a pinboard.

- In a second step, managers used these cards to enter a discussion process. For example, one manager stated 'We want to keep the positive culture and achieve continuity'.

- As an outcome of this discussion process, the five managers reached an agreement and formulated a shared vision for employees: 'We want to invest in employees, achieve continuity, quality and flexibility'.

While the example refers to the overall vision for employees, a formulation of a technology vision can be performed in a similar way.

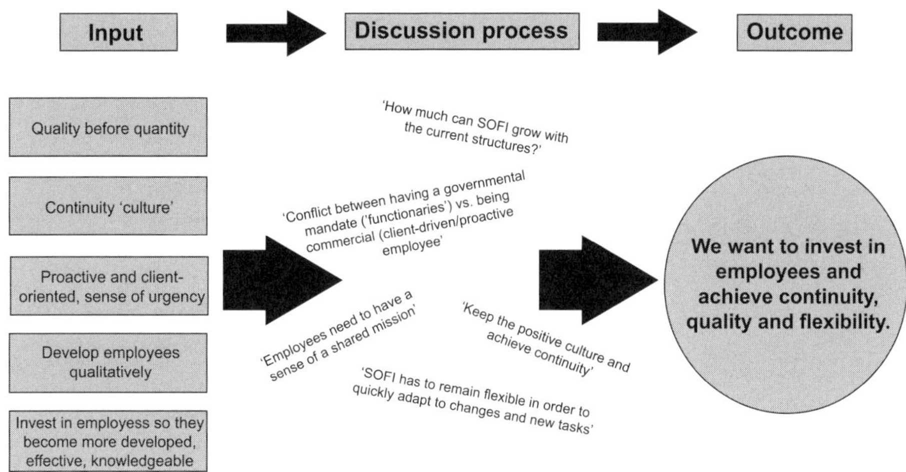

Figure 8.4   An example of the process of vision formulation

8.3 Strategic Analysis

The technology mission statement formulates a first anchor for the technology strategy. As a second step, a *strategic analysis* creates transparency regarding actual and future influencing forces on technology development. The following factors should be analysed (see Figure 8.5):

- *Internal factors* consist of the strengths and weakness of a company, region or country regarding technological advancement. Governmental organizations or investment promotion agencies can to some degree influence these internal factors. Factors are natural resources, human resources, financial resources, the existing prior technology base (see Section 4.2), communication structures (see Section 4.2), cultural characteristics, environmental factors and local laws and policies.

- *External factors* formulate threats and opportunities for technological developments. Governmental organizations or investment promotion agencies cannot directly influence these factors but rather are influenced by them. Factors include domestic and foreign markets for technology, international laws and policies, global technological trends as well as other regions and countries competing for technologies.

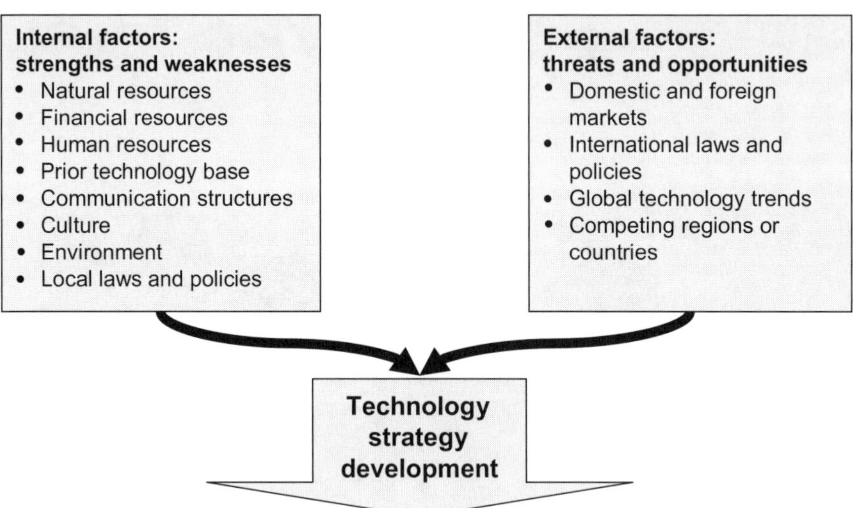

Figure 8.5   Internal and external influences

We will now describe each of the influencing factors briefly. Afterwards, we provide you with a simple method for their systematic analysis. We start with the internal factors.

8.3.1 ANALYSIS OF INTERNAL FACTORS (STRENGTHS AND WEAKNESSES)

A first field of analysis consists of the *natural resources* of a country or region. Natural resources may be classified into renewable and non-renewable resources:

- Renewable resources are generally living resources (fish, coffee and forests, for example), which can restock (renew) themselves at approximately the rate at which they are extracted, if they are not over-harvested.

- Non-renewable natural resources include mineral resources, such as oil, coal, copper, or gold.

As many production processes in developing countries are related to the utilization of natural resources, the formulation of technology strategy needs to take these into account. For example, technologies of offshore fishing are not well-suited for countries with no direct access to the open sea, such as Bolivia or Afghanistan. Also, oil refining technologies are predominantly of interest for countries disposing of oil reserves. Similarly, technologies related to tropical fruit production are of interest for countries such as Ecuador, Costa Rica or Brazil but not for nations with a continental climate such as the UK or the Netherlands.

Subsequently, the *human resources* of a region/country influence strategy formulation. While issues such as labour cost, unionism or labour codes are important for FDI in general, technology transfer depends to a large degree on the available skills and educational standards of a country. For example, India has become successful in attracting IT services and technology because it has internationally renowned universities in the field of software development. On the other hand, trying to attract high-end airplane technology in a country with few engineers in the field of aerodynamics will not be very successful. Managers can use a country's number of university graduates, percentage of primary and secondary school graduates, illiteracy rate, or university rankings as indicators for strengths and weaknesses in comparison to competing countries/regions.

In addition, the *financial resources* of an investment promotion agency or similar organization define the economic restrictions for technology strategy. Greater availability of financial resources allows governments or investment promotion agencies to formulate tax incentives or to invest in country promotion in order to attract foreign technology via FDI.

Another factor influencing the formulation of technology strategies is the *prior technology base* of an industry, region, or country. As we saw in Chapter 4, the prior technology base increases the ability to assimilate and use new technologies. For example, a country interested in attracting electronics technology should develop a proper understanding of its existing prior technology base in this field. Let's assume its firms are already in possession of assembling capabilities in the electronics sector (see Section 2.5). Then it would make sense to attract manufacturing capabilities for some of the assembled components in order to enhance the value chain (see the case study on Thailand's electronics industry in Section 12.3).

Further, the *communication structures* of a region or country should be analysed. As we saw in Chapter 4, a region's capacity to absorb new technology depends on the individuals that stand at the interface with potential providers of technology. If no sufficient communication channels exist between a nation and providers of technology, a technology may never be absorbed. Indicators such as membership in international organizations, industry associations or networks to technology holders and MNCs can be used to describe the status quo.

Following this, *environmental factors* require assessment before technology targets can be formulated. For instance, the quality of water, air, or soil may significantly affect the application of some technologies. Also, some technologies may involve a negative environmental impact that needs to be considered.

In addition, *cultural factors* exert an influence on technology transfer. The word culture generally refers to patterns of human activity and the symbolic structures that give such activity significance. Some technologies may be incongruous with certain types of cultures. For example, some technologies of food production cannot be used in certain countries because they are inconsistent with religious norms. Also, some cultures are more open to new technological developments than others. For instance, Japanese consumers are known for their willingness to experiment with new technologies in the field of mobile communication.

Finally *local laws and policies* need to be analysed before a strategy can be formulated. Laws and policies steering technology transfer may be tax incentives for certain industries, joint venture requirements, or import tariffs. Also, property rights play an important role. For instance, MNCs usually require some degree of property rights protection before establishing R&D facilities (see Section 11.2.3 for details).

How can managers analyse each of these internal factors in practice? We propose an analysis along the following lines:

- A neutral description of the *actual situation* for each influence factor, including expected *future developments*.

- Based on this, an analysis of the *strengths and weaknesses* of a region or country with regard to each factor. *Strengths* are characteristics of a region or country that are helpful to achieve the technology vision or mission. *Weaknesses* are characteristics harmful to the achievement of vision or mission.

Managers can perform such an analysis in a workshop or in interviews with experts. Table 8.4 can be used to structure the discussion and analysis process. Besides the influencing factors already mentioned above, more issues can be added if required.

Table 8.1   Analysing internal factors influencing technology transfer

Internal factors	Actual situation and expected future developments	Strengths	Weaknesses
Natural resources			
Financial resources			
Human resources			
Prior technology basis			
Communication structures			
Environment			
Culture			
Local laws and policies			
Other factors			

8.3.2 ANALYSIS OF EXTERNAL FACTORS (OPPORTUNITIES AND THREATS)

In addition to internal factors, *external factors* have to be considered when formulating a technology strategy. External factors formulate threats and opportunities for regions and countries.

First, *domestic and foreign markets* for certain technologies should be assessed. Markets are characterized by their size and growth, the prices for products or services, entry barriers and the number and behaviour of competitors. Managers should analyse domestic or foreign markets, depending on weather the product or process technology is focused on exporting or on production for the local market. Possible sources of information are government statistics, specialized market survey consultants, friendly firms already active in the market, chambers of commerce, the Internet, or others.

Secondly, *international laws and policies* influence technology transfer. Consequently, they need to be analysed before strategic themes for technology transfer can be formulated.

Thirdly, *global technological trends* should be considered. Here, the likely path of change of key technologies needs to be determined. Some key questions to be answered are:

- At what stage of the S-curve are technologies of interest positioned? (see Section 2.6)

- Are the technologies of interest still in the innovative stage or has a dominant design already emerged (see Section 2.6)?

Finally, managers have to take into account *other regions or countries* competing for technology transfer and FDI in the world market. For reduction of complexity, regions may be clustered into groups. Each group should consist of members with similar characteristics. Groups of regions or individual regions may then be studied with regard to the following criteria (see also Grant 1991: 82):

- Goals: what technology goals does the region have?

- Strategy: what is the region's strategy to acquire technology?

- Capabilities: what capabilities does the region have to realize the strategy?

Table 8.2   Analysing external factors influencing technology transfer

External factors	Actual situation and expected future developments	Opportunities	Threats
Domestic and foreign markets			
International laws and policies			
Global technology trends			
Competing regions or countries			

External factors can be analysed using the following criteria (see Table 8.2):

- a neutral description of the *situation* for each factor, including expected future developments

- derived from this, an analysis of the *opportunities and threats* for a region or country. *Opportunities* are external conditions that are helpful to the achievement of technology vision while *threats* are harmful.

Table 8.2 will help you to structure your analysis.

8.4 Formulation of Strategic Themes

8.4.1 DEFINITION OF STRATEGIC THEMES

While the vision formulates a picture of the desirable future, the strategy of an organization defines the logic through which this future may be achieved (Kaplan and Norton 2001: 74). A strategy consists of different *strategic themes*. 'In our experience, executives almost always separate their strategies into several focused themes ... In general, strategic themes reflect what the management team believes must be done to succeed' (Kaplan and Norton 2001: 78). Each strategic theme can be seen as a chain of hypotheses on cause–effect relations (Kaplan and Norton 2001: 78 f.).

As with the mission statement, '*technology strategy* is only one element of overall competitive strategy, and must be consistent with and reinforced by choices in other value activities' (Porter 1985: 176). Themes of technology strategy may address such topics as what technologies to acquire and to what degree to invest in a technology. 'The cost of improving the technology must be balanced against the benefit, as well as the likelihood that the improvement can be achieved' (Porter 1985: 179).

Also, technology strategy must deal with the development of technological infrastructure, such as the development of human resources and the promotion of indigenous R&D by domestic industry and research institutes (Cohen 2004: 222).

8.4.2 DEVELOPMENT OF STRATEGIC THEMES VIA SWOT ANALYSIS

How can managers integrate internal strengths and weaknesses and external opportunities and threats to formulate strategic themes for technology transfer? A reliable method is SWOT analysis. SWOT analysis combines internal and external factors in order to create different strategic options.

Table 8.3   SWOT analysis

External factors / Internal factors	Opportunities	Threats
Strengths	SO strategies	ST strategies
Weaknesses	WO strategies	WT strategies

To perform a SWOT analysis, the most important strengths/weaknesses and opportunities/threats are written down on the x and y axis of a matrix (see Table 8.3). Internal and external factors are then related to each other to generate strategic themes.

These are clustered into four categories (see also Müller-Stewens and Lechner 2005: 225):

- *SO strategies* draw on the internal strengths of a region/country to exploit the opportunities of the external environment. Typical are strategies that are aimed at expanding existing technologies or acquiring related technological knowledge. An example is Brazil's focus on ethanol fuel technology. On the one hand, Brazil is levering its strength of having a long history with ethanol technology and the required soil and climate for sugar cane production. On the other hand, it takes the opportunity of increasing energy and petroleum prices to capitalize on this technology.

- *ST strategies* aim to neutralize or moderate external threats through the use of internal strengths. Germany's heavy investments in wind turbine technology are an example of such a strategy. Germany faced the external threat of becoming overly dependent on foreign energy sources, especially on Russian natural gas. In order to reduce this dependency, Germany leveraged on its strong engineering skills and access to the windy North Sea to expand its wind turban capacity, partly substituting for the import of gas.

- *WO strategies* try to take advantage from opportunities of external factors to minimize internal weaknesses. For example, Singapore has taken advantage of a growing importance of private banking for wealthy clients (opportunity) and invested in banking technology. As a result, it has minimized its dependency on the industrial sector which was suffering from Singapore's lack of natural resources (weakness).

- *WT strategies* want to minimize threats of external forces through the reduction of internal weaknesses. Because the combination of external threats and internal weaknesses is the most dangerous combination for a region/country, this strategy is often pursued by organizations.

When formulating strategic themes, managers need to take account of the overall economic strategy of their country. For example, it makes a difference for the technology strategy if the national economic strategy of a country is built on market-oriented, laissez-faire premise or consists of a highly planned framework with centralized control (Cohen 2004: 223).

8.4.3 ORGANIZATION OF STRATEGIC THEMES

After developing strategic themes with SWOT analysis, managers can organize these with the help of strategy maps (Kaplan and Norton 2001). To formulate strategy maps, one can pursue the following steps:

- At the beginning, all strategic themes are written down on a pinboard.

- Subsequently, strategic themes are organized by analysing their relationships in terms of cause–effect relations and intensity. Strategic themes may be contradicting or enforcing each other. Symbols such as '++' or '-' may be used to illustrate relationships and intensity.

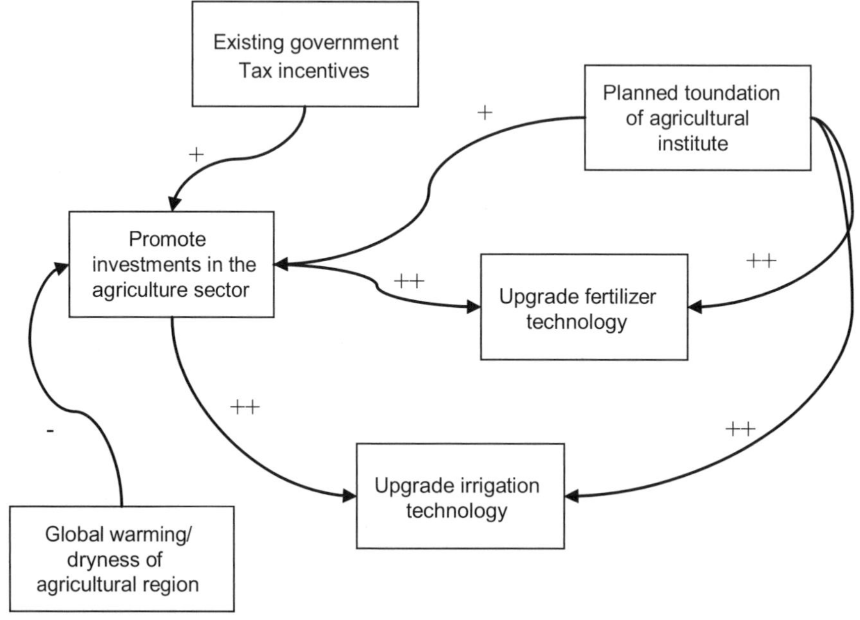

Figure 8.6   An example of a strategy map

Figure 8.6 gives an example of a strategy map for promoting investments in the agricultural sector.

8.5 Technology Strategy Implementation

The *implementation of technology strategy* refers to the development and execution of concrete measures. With the strategy map, managers can formulate measures for the most relevant strategic themes. Resource and time restrictions have to be considered. For each strategic theme selected for realization, managers can develop measures for the following categories:

- *Identification of technologies*: what measure can be used to identify the technologies of interest in the world market? How can we encounter potential sources of technology and technology providers?

- *Assess technologies*: what actions are necessary to evaluate identified technologies?

- *Attraction of technology*: what measures support the attraction of technologies defined in our strategic themes?

- *Absorption of technology*: what needs to be done to support the process of technology absorption?

- *Application of technology*: what measures are suited to facilitate the application of strategic technologies?

- *Performance measurement*: how do we monitor the whole process?

As you can see, the fields of action are structured by the phases of the technology management process introduced in Chapter 7. In addition to measures, performance measurement criteria may be formulated. Managers can develop both in a workshop with the help of Table 8.4.

Table 8.4   Development of measures for strategic themes

Strategic theme		
Field of action	**Measures**	**Performance criteria**
Identification of technology		
Assessment of technology		
Attraction of technology		
Absorption of technology		
Application of technology		
Performance measurement		

We will describe detailed potential measures for each phase of the technology management process in Chapters 9 to 14.

It is noteworthy that political factors are likely to exert a great influence on the described strategy process. 'Professional planners are frequently frustrated by what they often regard as "political interference" and by their inability to achieve meaningful results because of the nature and distribution of socio-economic and political power' (Cohen 2004: 219). In many cases a technology strategy might be limited by the political environment of countries. Therefore it is important that political influences are already considered during the process of strategy formulation and anticipating measures are defined.

8.6 Strategy Example

We now want to illustrate how each step of the strategy process can be carried out. We use a fictive case study on Brazil. The aim of the case study is not to be authentic in terms of the characteristics and technology strategy of Brazil. Instead, it serves to demonstrate the interaction of the different concepts provided above.

Having a population of over 180 million and being the biggest South American country in size and economy, the analysis provided remains superficial. For an authentic scenario, an in-depth analysis including an extensive data analysis would have to be executed.

The fictive project to develop a technology strategy for Brazil is divided into the following steps:

- Step 1: formulation of technology mission statement.

- Step 2: strategic analysis.

- Step 3: formulation of strategic themes.

- Step 4: strategy implementation.

We will describe how each of the four steps could be carried out.

8.6.1 STEP 1: FORMULATION OF A TECHNOLOGY MISSION STATEMENT

As a first step, the top management of the Brazilian Investment Promotion Agency holds a workshop. The aim is to develop a technology mission statement. Managers are asked to write their ideas for a technology mission, vision and central values on cards. Afterwards, these cards are put on a pinboard and discussed. Such a process could result in the mission statement shown in Table 8.5.

Table 8.5   Technology mission statement for Brazil (fictive)

Technology mission	To promote and stimulate the scientific and technological development of our country.
Values	Transparency, business driven and service-minded.
Technology vision	Establish Brazil as the major technological hub in Latin America.

8.6.2 STEP 2: STRATEGIC ANALYSIS

In a second step, an analysis of internal and external influencing factors is carried out. In a workshop, managers work on a pinboard to develop a picture for each factor.

Table 8.6 shows the results for the internal factors. For example, managers see a strength in the great diversity of natural resources and the already existing commercial use of these. Also, the human resources situation is perceived as twofold. On the one hand, the country is in possession of good universities (strength). On the other hand, there is a large amount of the population with low-level education (weakness). In addition, Brazil's open culture is seen as a major strength of the country.

Table 8.7 contains the threats and opportunities of the external environment. For instance, managers perceive the large domestic market of Brazil and the increasing per capita income as an opportunity for the country. In addition, the emerging markets of MercoSur are perceived as a potential destination for exports.

Table 8.6   Internal factors influencing technology transfer (example)

Internal factors	Situation	Strengths	Weaknesses
Natural resources	Almost all kinds of natural resources, for example minerals (iron ore, bauxite), oil, water, forest, etc.	• Great diversity of resources • Commercial use already taking place in most sectors	• In some sectors use of renewable resources is not sustainable, for example excessive deforestation in the Amazon region
Financial resources	Tax system guarantees financial resources for strategy implementation	Sufficient financial resources available	
Human resources	• Great amount of labour available • Large amount of population with low-level education • Good universities and higher education system • Governmental effort to reduce amount of non-educated people	• Good universities and higher education system	• Large amount of population with low-level education • Relatively poor foreign language education
Prior technology basis	• Newly industrialized country • Broad technological base across all industries, for example automobile, airplane industry, services	• Technological base is diversified across all industries • Technology clusters serve as a catalyst for R&D	• R&D activity needs to be increased • Information infrastructure (broadband Internet, etc.) needs to be advanced
Communication structures	• Sufficient communication structures via various governmental agencies at national and state-level • Industry maintains open channels to foreign technology providers.	Sufficient communication structures available on public and private level	
Environment	• Environmental damage to some degree, e.g. deforestation, water pollution • Laws and legislation regulating environmental pollution	Laws and legislation regulating environmental pollution	Enforcement of environmental laws could be improved
Culture	• Cultural mix with European and African elements • Open culture	Open culture	
Local laws and policies	Some protectionism regarding imports	• Solid political institutions • Incentives for FDI because of import tariffs	• Test of new foreign products via imports is difficult • Some import tariffs restrict entrance of firms in domestic market and lower incentives for technological advancement of domestic firms • Long-lasting trials

Table 8.7   External factors influencing technology transfer (example)

External factors	Situation	Opportunities	Threats
Domestic and foreign markets	• Very large domestic market • Increasing per-capita income • South–South commerce, for example with China for natural resources • Emerging market of MercoSur	• Very large domestic market • Increasing per-capita income • South–South commerce • Emerging market of MercoSur	
International laws	—		
Competing regions or countries	• In South America almost none (except for certain industry sectors) • In Latin America: Mexico • India with regard to pharmaceuticals • China, Russia, Egypt in automotive industry • Relatively high production cost in comparison with Asia, CIF	• In South America almost none (except for certain industry sectors)	• In Latin America: Mexico • India with regard to in pharmaceuticals • China, Russia, Egypt in automotive industry • Relatively high production cost in comparison with Asia
Global technology trends	Growing demand for information technologies		Language isolation raises costs for implementation of information technologies

8.6.3 STEP 3: FORMULATION OF STRATEGIC THEMES

In the next phase, the information of the strategic analysis is used to develop strategic themes. Firstly, a SWOT-Analysis is performed to generate different strategic themes. To do so, the most important strengths/weaknesses and opportunities/threats are written down in a matrix such as Table 8.8.

Table 8.8   SWOT analysis (example)

External factors / Internal factors	Opportunities	Threats
	• Very large domestic market • Emerging market of MercoSur	• Relatively high production cost in comparison with Asia, CIF countries • Increasing competition from China, Russia, Egypt with regard to automotive industry
Strengths • Good universities and higher education system • Technological base is diversified via various industries • Technology clusters serve as catalysts for research and development and attract FDI	**SO strategies** • Use good human resources and large domestic market to attract new technologies in all industries	**ST strategies** • Use good human resources to attract new technologies in automotive industry • Technologically outrun competitors in automotive industry to compete via quality not price • Use regional technology clusters to attract R&D in the field of automotive industry
Weaknesses • Large amount of population with low-level education • Information infrastructure needs to be advanced	**WO strategies** • Use domestic market size to attract FDI with information technology (next generation mobile phone, Internet)	**WT strategies** • Increase productivity by improving second degree education

In the workshop, the management agrees to pursue an ST strategy, which means neutralizing or moderating external threats through the use of internal strengths. The strategy wants to increase productivity in the automotive industry in order to legitimize higher wages in comparison with competing countries. Technology clusters and the relatively good human resource base will be used to achieve this. For a more complex analysis, strategic themes could be organized in a strategy map.

8.6.4 STEP 4: TECHNOLOGY STRATEGY IMPLEMENTATION

In order to pursue the developed strategy, implementation measures have to be developed. Table 8.9 gives an example how measures for the ST strategy could look.

Table 8.9   Development of measures (example)

Strategic theme: attract R&D in the automotive industry		
Field of action	**Measures**	**Performance criteria**
Identification of technology	• Intensify presence at automotive fairs	• Fairs visited
	• Build up contacts to R&D departments of international automotive companies	• Contacts established
Assessment of technology	• Apply DCF analysis	• Discounted cash flow value
	• Conduct a Delphi survey on future of automotive industry	• Delphi results
Attraction of technology	• Take policy advocacy role for tax allowances for R&D in automotive industry	• Establish contacts with journalists • Establish contacts with members of parliament • Meeting with economic minister
Absorption of technology	• Foster development of automotive science parks	• Meetings with governor
	• Promote local automotive suppliers to facilitate vertical spillovers	• Matchmaking event with local suppliers and international automotive companies
	• Foster the foundation of joint ventures	• Promote local joint venture partners
Application of technology	• Support international marketing of technologies acquired by suppliers	• Perform annual marketing seminar
Performance measurement	• Develop performance measurement system	• Formulation of a performance measurement process

9

Gathering Information on Technologies

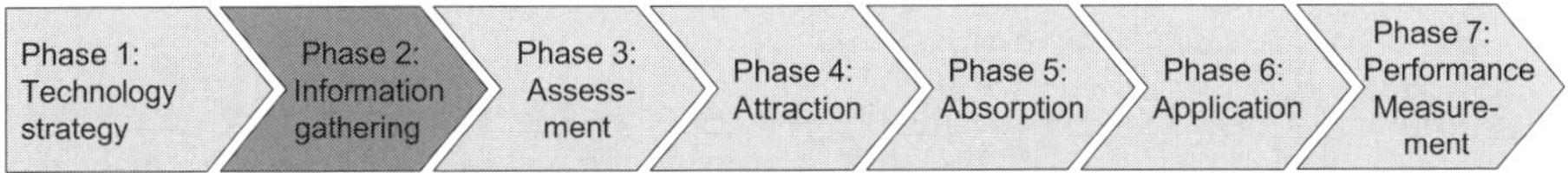

Figure 9.1   Information gathering (phase 2)

In the first phase of the process of managing technology transfer we have explained how a technology strategy can be developed (Chapter 8). During this development, managers decide which category of technology a country should target for its development. In the second phase of managing technology transfer, managers focus on the gathering of information on these technologies (Figure 9.1).

Evidence shows the improper transfer of technology to developing countries causing many social, cultural, political, environmental, regional and economic problems (Cohen 2004: 231). For example, Thailand has made great technological advances that have stimulated economic growth over the last 50 years. However, the introduction of new technologies has also caused deforestation, pollution, degradation of water courses and waste production. Therefore, it is important to collect information about the available technology options (Cohen 2004: 231). A decision on which technologies to transfer is, at most, as good as the options evaluated. However, acquiring information about technologies is related to certain problems. To identify a suitable technology, an organization has to fully understand its content and functioning. Once it is capable of this, no incentives remain to pay the technology provider because the organization can now duplicate or copy the technology itself (see Section 2.2). Technology providers have strong incentives to guard information about their technologies but this complicates the process of information acquisition on technologies.

To be realistic, especially for developing countries, it will not be possible to be wholly informed about all technological options (Cohen 2004: 231). Still, some possibilities exist to obtain information on technologies before actual acquiring them. In this chapter we want to introduce you to these possibilities:

- we begin this chapter by describing some categories of technology on which information can be obtained (Section 9.1)

- you will then get to know some sources of information about technologies (Section 9.2)

- we conclude by introducing two different searching techniques (Section 9.3).

9.1 Categories of Information on Technologies

Information on technology is required to prepare for decisions as to whether to acquire a certain technology or not. Information on the following dimensions should be gathered:

- *Capability of technology.* In Chapter 3 we explained that technology is generally used to transform certain inputs into outputs. One should therefore analyse the capabilities and benefits of a technology and the outputs it can produce. Let's take the balanced scorecard, a management technology for strategy implementation, as an example. Its capabilities are the translation of strategy into measurable parameters, the communication of the strategy to all employees and the feedback of strategy implementation.

- *Components of technology.* Often technology is not an integrated unit but may consist of different components. For example, mobile phone technology consists of technology for mobile phone towers, mobile phones, software protocols and managerial capabilities. To analyse a technology, information on components ought to be collected.

- *Stage of technological evolution.* Before acquiring a technology, one should acquire information on the stage of evolution of a technology. Such information is important as investing in a technology that is still in the innovative stage involves the risk that this technology could become replaced by another dominant design (see Section 2.6). Also, the question of whether a technology is already at the end of its life cycle, risking it being replaced by another innovation, should be considered.

- *Variations of technology.* Many technologies exist in different forms. For example, automobiles may be run with gasoline, electricity, solar, or other forms of engines. If different forms exist, data has to be gathered on similar forms of technology and its characteristics.

- *Technology protection.* Information should be acquired on possible measures protecting the technology from being transferred. Examples of such measures are international patents or technology standards of countries.

- *Technology providers.* Once the capabilities of a technology and its components are determined, providers of technology need to be identified. These may be one or various private companies, government agencies, or international organizations. Information is necessary on the reputation, financial situation, experiences, technological capabilities, strategy and interests of possible providers. Let's take the automotive industry as an example. An investment promotion agency wanting to attract technology in this field should try to understand the global supply strategy of a manufacturer, its financial position, its willingness to transfer technology to local suppliers and the state of its technology (see Section 5.4.)

- *Transfer intermediaries.* Developing countries should also seek to identify possible facilitators of transfer for the technology. These may consist of multilateral organizations or business councils capable of establishing contacts to providers and of moderating the transfer process.

- *Market information.* Information has to be acquired on the market potential of the technology under analysis. This includes information on domestic and foreign market potential, consumer habits and fashion, market growth estimates, main competitors and possible substitutions for the technology.

- *Possible socio-environmental impact.* To estimate the possible impact of new technologies, it is useful to collect information on influences on the socio-environmental characteristics of a country. Historical data may be used from similar countries to which the technology has already been introduced.

- *Cost of technology.* Usually providers of technology have invested in technological development and therefore require financial or some other form of compensation. Costs of technologies or the costs of incentives to attract these have to be estimated in order to calculate a cost–benefit ratio.

- *Explicit vs. tacit technology.* As discussed in Section 2.3.2, technologies can exist in explicit or tacit form. Both forms require different strategies for transfer and need distinct amounts of resources (see Section 2.3.2).

- *Medium of technology storage.* Finally, information on the medium of technology storage should be gathered. While the transfer of hard mediums might be uncomplicated, the transfer of technologies stored in human and collective mediums proves to be more complex. Labour might have to be contracted or business consultants hired.

9.2 Sources of Information on Technologies

A variety of sources to acquire the necessary information exists (see also UNIDO 1995):

- *Patent offices*: patent information is relatively recent in terms of technology information. Because a wide range of information has been accumulated over a long period of time, it can be the priority search target in the technology discovery process. Above all, patent information may be acquired to find owners of technology. Patent Offices such as the European Patent Office or the United States Patent and Trademark Office have websites that can be searched.

- *Scientific journals and books*: scientific journals and books provide access to recent scientific knowledge on technology. With subscription, many journals may be accessed directly over the Internet and provide full search capabilities.

- *Universities and public research institutes*: although R&D is now to a large degree performed by private firms, universities and public research institutes still provide a reliable source for information on technological developments. University websites contain extensive information on scientific projects and experiences of staff.

- *Trade fairs and shows*: a trade fair is an exhibition organized so that companies in a specific industry can demonstrate their new products and services. Trade fairs offer good opportunities to acquire information on technologies and technology providers. They also provide the chance for personal contacts with potential technology suppliers.

- *Engineering bureaus*: engineering bureaus play a critical role as their activities range from simple consulting and advice to the transfer of complex technology, including the supply of turnkey plants. Some of them acquire technology from the owner under contract, add their design, know-how and engineering and sell the complete package, including the grant of license. Others focus only on contracting and licensing of technology (UNIDO 1995: 10).

- *Finished products*: finished products allow insights into technology characteristics. Capabilities of certain technologies as well as design features can be identified.

- *Technology transfer intermediaries*: various national and international organizations try to facilitate the transfer of technology and provide a match between technology providers and demanders. For example, the Asian and Pacific Centre for Transfer of Technology (APCTT) aims to facilitate technology transfer to the Asia-Pacific region (see Box 9.1).

- *Private companies*: a large degree of R&D is today performed by private firms. In particular, multinational corporations in knowledge-intensive industries invest heavily in research and development. Company websites may provide you with some information on technologies in possession of private companies. For stock listed companies, investor relations documents such as the annual report or financial analyst presentations are an additional source of information.

- *The internet*: the Internet provides easy to use search engines to find information on technologies.

- *Social networks*: the importance on personal and social networks for technological information cannot be underestimated (see Section 4.3). In fact, much information on technology, by the time it is published, is already old news. A private company, investment promotion agency or government body should therefore seek to establish a wide network with external parties (see also Section 4.4). To do so, organizations need to encourage their employees to engage in boundary-spanning activities such as attending conferences, relate to universities, visiting companies and actively participating in industry and trade associations (see also Shane 2009: 341).

BOX 9.1   THE TECHNOLOGY MARKET OF APCTT

The APCTT is a regional institution of the United Nations Economic and Social Commission for Asia and the Pacific (UNESCAP). Its objective is to facilitate technology transfer in the Asia-Pacific region. The Centre is headquartered in New Delhi and provides services such as information on technology/investment opportunities, matching of business partners and support services (market/ feasibility studies, technology evaluation, contract negotiation). The website of APCTT also contains TechMart, an interactive technology market database. The database focuses on environmentally sustainable technologies and invites individual technology suppliers and seekers to submit technology offers and requests in a prescribed format. Each database entry consist of a short description of the technology, its area of application, its advantages, information about environmental aspects, the development status and the transfer terms.

Source: http://www.apctt.org

9.3 Searching Techniques

Two different searching techniques may be combined to acquire information on the categories mentioned in Section 9.1 (see Müller-Stewens and Lechner 2005: 208):

- *Systematic scanning* refers to an open search process for new technologies and trends. Like a 360-degree radar the world market is searched in a relatively open manner. A great variety of sources is used to provide information. Due to the great variety of technologies analysed, the depth of information on each technology is relatively low.

- The *focused search* describes the monitoring of single technologies. Extensive information is gathered on selected technologies or technological trends. Specific sources are used to formulate a dense profile of the technology of interest. Information depth on the analysed technologies is high.

Both methods should be combined to avoid overseeing important technologies and at the same time to gain the necessary depth of information.

Technology Assessment

10

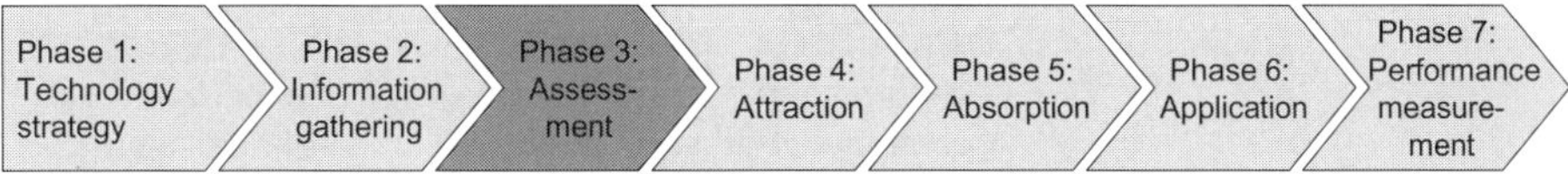

Figure 10.1   Technology assessment (phase 3)

In phase 2 of the process of managing technology transfer, we saw how information on technologies can be gathered (Chapter 9). In phase 3, we want to illustrate how this information can be used in order to assess technologies. Assessing a technology means comparing it with other options and asking if a given technology will do better, cheaper, or more easily than other technologies (see also Braun 1998). We also need to ask what impacts a technology might have for a country and what side effects it might provoke. To provide you with the required concepts and tools for technology assessment we have structured this chapter as follows:

- First, we describe the process of technology assessment and the questions that need to be answered (see Section 10.1).

- We then introduce you to the different methodologies and decision-making tools of technological assessment (see Section 10.2).

- Finally, we demonstrate best practices for technology assessment (see Section 10.3).

10.1 Phases of Technology Assessment

Step 1: Defining the scope of the assessment	Step 2: Describing technology	Step 3: Assessing benefits and costs	Step 4: Addressing unwanted effects
Key questions What is the scope of assessment?	• What are the characteristics of the technology? • What are its alternatives/complementaries? • What are the likely development paths?	• What benefits and costs can be expected from the technology? • To whom will these benefits and costs accrue? • Does the technology fit with the technology strategy?	What dangers does the technology harbour and what ill-effects might it cause?

Figure 10.2   Steps in technology assessment

The process of technological assessment consists of four different steps (see Figure 10.2) (Braun 1998: 34–38, Phillips 2001: 243):

- First, one needs to define the *scope of assessment*. Do we want to analyse a narrowly defined technology and its immediate rivals, or do we want to assess a major bundle of technologies serving related purposes? For example, will we evaluate fibre optics or information technology? If chosen too narrowly, the assessment will be unlikely to reveal anything of great interest. If chosen too widely, the evaluation will become very expensive, complex and is unlikely be of much practical help for the decision-making process.

- The second step of technology assessment consists of a *description of the technology* under scrutiny. This step is based on the information gathered in phase 2 (Chapter 9). Performance, principals of operation and costs are regarded as major descriptors of technologies. A description should also make reference to complementary and alternative technologies. *Complementary technologies* make a technology feasible and practical, or more efficient and wider in scope. For example, the scope of the Internet is extended via online payment technologies. *Rival or alternative technologies* fulfil much of the same function as the technology under discussion and may substitute it. For example, satellite TV broadcasting technology cannot be discussed without mentioning the rivalry with cable television.

- The third step of technology assessment concentrates on the core questions: what *benefits and costs* are to be expected from the technology, what needs does it satisfy and why is it superior to rival technologies? The benefits of a technology may be purely economic and commercial, or they may be expressible in terms of environmental improvements, health benefits or improvements in the social or political circumstances. As benefits in general involve some value judgement, one should analyse non-controversial benefits as well as those impacts that might be regarded as benefits by some and not by others. It is also necessary to show whom the benefits might accrue to. Also, costs of a technology, for example in the form of tax incentives or licence fees, should be discussed. In addition, one needs to discuss the fit of the technology with the technology strategy of a country or region.

- Fourthly, *unwanted effects* or hazards of a technology should be addressed. In case of undesirable effects, the question of who or what might be adversely affected should be addressed. As in the third step, it might be controversial to decide whether some effects of a technology are 'good' or 'bad'. In any case, possible side effects which prove destructive of the natural environment, dangerous to human health, disruptive of society, or otherwise trigger a chain of events which appears unpredictable and risky, have to be mentioned.

Box 10.1 demonstrates the process of technology assessment for solar energy technology.

BOX 10.1   THE PROCESS OF EVALUATING SOLAR ENERGY TECHNOLOGY

Imagine conducting a preliminary technology assessment of solar energy technology for a developing country. This would require the following steps:

Step 1 (*scope of assessment*): should solar energy technology be assessed in general? Or shall we focus on more specific sets of technology such as solar thermal electric power plants, photovoltaic cells, or solar heating? Or, more narrowly, should a single technology such as a solar thermal power plant or power towers be evaluated? In this example, we focus quite narrowly on a solar thermal power plant.

Step 2 (*description of technology*): the solar thermal power plant needs to be described. In a solar thermal power plant oil in the receiver tubes collects the concentrated solar energy as heat and is pumped to a power block for generating electricity. Rival technologies consist of other forms of energy production such as water power or nuclear power plants. More substitutes are other solar energy technologies such as a solar dish engine or a power tower. For detailed analysis, performance measures and costs of each option have to be analysed.

Step 3 (*assessing benefits*): benefits of a solar thermal power plant include:

- It is relatively pollution free.
- Once facilities have been constructed, they run with little extra input or maintenance.
- There is no dependence on foreign energy suppliers.

Some disadvantages compared to traditional forms of energy production are:

- It is only available in certain areas of favourable climate and latitude. These locations tend to be remote from the places of highest energy demand.
- It is not available at night and is reduced when there is cloud cover.
- It must be converted into some other form of energy to be stored.

Step 4 (*unwanted effects*): although running quite pollution free, the plant may cause environmental damage during manufacturing and construction.

10.2 Methods of Technology Assessment

In order to forecast the benefits and costs of a technology (step 3 in Figure 10.2) as well as its unwanted effects (step 4 in Figure 10.2), a variety of methods can be used. The following methods support you in performing a systematic assessment:

- *Trend line analysis* (Section 10.2.1) and *modelling* (Section 10.2.2) allow you to forecast certain key variables of technologies, such as the demand for technologies, price developments of technologies or associated costs.

- *Scenarios* (Section 10.2.3) develop several possible futures related to technologies.

- The *Delphi method* (Section 10.2.4) uses experts' opinions to gain insight into future developments of technologies.

- *Discounted cash flow analysis* (Section 10.2.5) allows to calculate the financial return of a technology investment project.

- *Cross-impact analysis* (Section 10.2.6) is aimed at identifying unintended consequences of technologies.

- Finally, *utility analysis* (Section 10.2.7) supports the selection of technological alternatives or the choice between different sources of technology.

10.2.1 TREND-LINE ANALYSIS

Trend-line analysis obtains historical data and fits a curve to it which extends into the future. Such an approach is based in the assumptions that things will continue as before. In order to formulate a trend-line, it is very important to obtain a good set of historical data covering a long period of time. For example, historical data might show that the demand for notebooks in a certain country has increased by 10 per cent every year over the last 15 years. Assuming this trend to continue, one can forecast the future demand for notebooks by simply extending the trend-line.

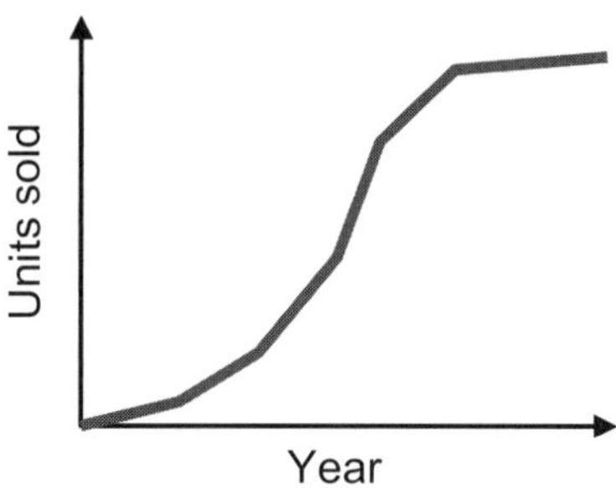

Figure 10.3   Market saturation

The main risk of this method is an unforeseen change in trend. A well-known example is the incorrect extrapolation of trends by not considering either *saturation or substitution* effects. The growth in sales of a certain product may follow a linear or even exponential curve for a while, but before long the market is saturated. The result is some form of S-curve (see Figure 10.3). While this effect is well known, the problem is to forecast at what time the S-curve will approach its asymptotic value and what this value will be (Braun 1998: 111).

While this may already be difficult, a change in trend caused by an unforeseen event will cause trend-line analysis to be irrelevant. Such an unforeseen event occurred in 1973 in the rise of the price of crude oil. It may also consist in the sudden realization that a used substance poses a major hazard to health, as was the case with asbestos. A recent prominent example of a change in trend is illustrated in Figure 10.4 and Box 10.2.

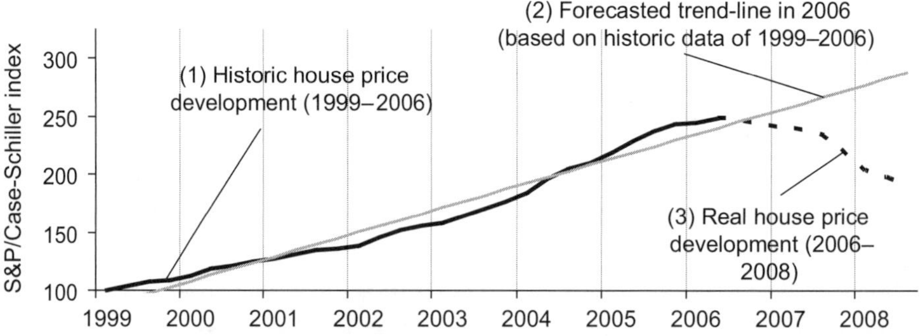

Figure 10.4   S&P/Case–Schiller Index of 10 metropolitan cities (USA)

BOX 10.2   TREND-LINE ANALYSIS AND THE US HOUSING CRISES OF 2007–2008

A well known example of an unforeseen break in trend is the US housing crises of 2007–2008. The collapse of the US housing market eventually resulted in the global financial crises. It provides a good example of how simply extrapolating historic data into the future can result in disastrous forecasts.

Figure 10.4 shows the development of the US house prices from 1999 to 2008. The following observations can be made (see numbers in Figure 10.4):

1. House prices have increased continuously from 1999 to 2006, accelerating from 100 to 250 index points.
2. Applying an extrapolation in 2006 with the help of a trend-line results in the incorrect conclusion that house prices will increase even further. However, such an assessment neglected the fact that house prices were already overvalued and a housing bubble had formed.
3. In reality, the extrapolated trend never materialized. Instead, house prices collapsed over the following years, resulting in one of the greatest devaluations of assets in economic history and in a major financial and economic crisis.

A special type of trend-line analysis is drawing conclusions from *analogies*. Analogies transfer experience from one field of technology implementation to another. As technologies transferred to developing countries have often already been used in other countries, this provides a useful method for forecasting. For example, conclusions may be drawn from the past introduction of solar energy technology to other countries because it is assumed that history will repeat itself. Similar to extrapolation in general, it needs to be carefully analysed if two countries or regions share sufficient characteristics. More sophisticated conclusions may be drawn if parallels exist in more than one country or region. However, two cases are never completely comparable. Although there may be a lot of parallels between two situations, no guarantee exists that the same effects of technology introduction can be observed.

In summary, trend-line analysis can give useful results provided managers have good historical data and provided they take an intelligent look at the cause of the trend and the effect of possible changes in circumstances (Braun 1998: 114).

10.2.2 MODELLING

Whereas trend-line analysis extends real historical data into the future, modelling is based on a mathematical formula capturing the relationships between two or more variables. In such a model, an independent variable, for example national income, may be used to predict a dependent variable such as demand for automobiles.

Two conditions must be fulfilled to obtain reliable results from models (Braun 1998: 121):

- there must be a causal relationship between the variables under analysis; and

- the relationships must hold over the time period of the forecast.

Well known examples of forecasting models are meteorological and economic models. For example, an economic model may use recent figures for variables such as prices, manufacturing output, balance of trade, money supply and interest rates in order to forecast economic growth.

Box 10.3 gives an example of a well known model.

BOX 10.3   THE MEADOWS' MODEL OF RESOURCE USE

One classical model that has achieved great prominence in technology forecasting is that published by Meadows and colleagues in 1972. The essence of the model is to show that if the use of natural resources continues to grow exponentially with the time, then the limited reserves will, sooner or later, be exhausted. Using the computer's capability of simultaneously dealing with a large number of variables, the model shows how, with different scenarios of use and conservation, the natural and economic system will face some form of collapse. The model is extremely complex and uses many non-linear relationships.

Although the model has been heavily criticized by some people, it was capable of showing that exponential growth is unsustainable in the long run and that various conservation methods can make a difference.

Adopted from Braun (1998: 121).

In order to forecast the development of a certain technology with a model, the following steps have to be performed:

- Step 1: you need to identify if there is a causal relationship between two variables, e.g. electricity consumption (E) and national income (I). This can be done with a correlation analysis, which can be performed in a spreadsheet program such as Microsoft Excel.

- Step 2: a formal relationship needs to be established between the two variables under analysis. Such a relationship can be calculated with a regression analysis. For example, you might come to the conclusion that electricity consumption (E) grows twice as fast as the national income (I). Then the relationship could be formulated as: $E = 2 \times I$.

- Step 3: the formula can now be used to forecast the development of the dependent variable. For example, if the government expects a growth in national income (I) of 2 per cent for the next year, according to the formula, electricity consumption should grow with 4 per cent (2×2 per cent).

The problem with a model is that social systems are extremely complex and often cannot be reduced to the relationships established in a model. Nevertheless, a model can often provide a first insight into potential future outcomes.

10.2.3 SCENARIOS

Scenario analysis is very different from extrapolation and modelling. It does not try to predict what will actually happen but rather develops several possible futures, typically two to four different scenarios (Mellahi et al. 2005: 90). Scenario creation is a way of envisaging what the future might hold for a particular economy, industry sector or organization. It follows a systematic sequence of steps (see APEC 2002: 16, Mellahi et al. 2005: 90, Shane 2009: 109):

- First the key drivers – social, technological, economic, environmental and political – are identified.

- Subsequently, key drivers are separated into *predetermined elements*, for example demographics, and *critical uncertainties*, for example public opinion or economic crises.

- Thirdly, forecasts are developed for predetermined elements and critical uncertainties.

- Thereafter, different forecasts on predetermined elements and critical uncertainties are bundled into scenarios. These are internally consistent stories which distinctly different possible futures.

- These scenarios are then interpreted and assessed. They are used to make strategic decisions and develop certain policies.

10.2.4 DELPHI METHOD

Another method for forecasting benefits, costs and unintended consequences of a technology is the so-called Delphi Method (Braun 1998: 115–120). It attempts to obtain a consensus of opinion among experts in a certain field. Experts are asked questions such as 'When will a reliable and effective artificial heart be available for transplantation into humans?'

The method involves several rounds of questions. The answers in the first round are analysed and the analysis is submitted to the same experts in an attempt to find consensus. In a second round, experts might be approached with a statement such as: '60 per cent of the experts suggested that the event in question will occur within ten years. In the light of this, would you like to modify your original estimate?'

The method is basically founded on two assumptions:

- experts working in a certain field have a good feel of how the field might progress and when certain key results will occur

- aggregated opinions are more reliable than single opinions of experts.

Expert opinion may also be used without the formalization required by the Delphi Method, for example only using a single round or conducting individual interviews with experts.

The Delphi method faces the problem that experts are busy people and often unwilling to give much of their time. Some experts are also unwilling to share their knowledge for distinctive reasons. In addition, Delphi results are highly sensitive to the formulation of the questions asked. Finally, when experts are asked to give their opinion on future events, the questioner should consider where their expertise lies.

Box 10.4 gives an example of the use of the Delphi method.

BOX 10.4   AN EXAMPLE OF THE DELPHI METHOD: FORECASTING WATER MANAGEMENT

In 1998, the APEC Center for Technology Foresight conducted a multi-country foresight study on water supply and management using the Delphi method. Experts and stakeholders across the APEC member economies were surveyed for their views on a series of topics. The table below presents a selection of the 64 topics surveyed, that more than two-thirds of respondents (in the second round) considered had a high degree of importance.

Topic	High importance (%)	Mean year of realization – APEC region	High need for APEC cooperation (%)
Technology to detect/locate leaks of over 10% from the distribution system developed	87.9	2005	43.6
Water pricing systems used to control demand	82.9	2007	23.1
70% of water used in industry recycled for further use	75.8	2009	33.3
Accurate rain and precipitation water-balance forecast, aiming at effective utilisation of rainfall in widespread use	74.3	2007	48.3
Nationally determined priorities of water usage/sharing among the sectors (domestic, industrial and agricultural) enforced	73.5	2006	22.6
Irrigation systems exceed 75% efficiency	72.7	2008	30.9
International standards for dam safety are enforced	72.2	2005	48.1

Adopted from APEC (1998).

10.2.5 DISCOUNTED CASH FLOW ANALYSIS

Discounted cash flow (DCF) analysis is a quantitative financial method for technology assessment. It calculates the benefits of a technology investment in terms of its financial return. DCF analysis uses free cash flow projections and discounts them (most often using the weighted average cost of capital) to arrive at their present value. If the discounted cash flow is higher than the current cost of the investment in technology, the investment opportunity may be a good one. The following formula is used to calculate the DCF:

$$DCF = \frac{CF_1}{(1+r)^1} + \frac{CF_2}{(1+r)^2} + ... + \frac{CF_n}{(1+r)^n}$$

CF = cash flow of period
r = discount rate

In order to calculate the free cash flows (CF), managers have to make a variety of assumptions on the development of product sales, prices and costs. Managers can use trend-line analysis or modelling to develop such assumptions.

In the context of DCF analysis, it is noteworthy that the costs of a technology are not necessarily borne by those who share the benefits of technology transfer. For example, the building of an oil pipeline in a developing country may cause benefits for employment, the oil company and the government on the one hand. On the other hand, costs may arise for the local population because of environmental damage and pollution.

In addition, two other problems of DCF analysis exist:

- Assigning financial values to intangible costs and benefits can become complicated. Imagine the construction of a high-speed train link between San Francisco and Los Angeles, two Californian cities: how can the value of saving two hours travelling time between these two cities be financially valued?

- Secondly, the outcome of DCF analysis is very sensitive to the assumptions taken. This is especially problematic as forecasts have to be developed over a very long period of time, increasing the probability of making false projections.

Nevertheless, DCF analysis is an important method in the financial assessment of any technological investment. However, it should be treated with caution when evaluating intangible costs and benefits. The financial value of the natural environment, ancient monuments, or human life is difficult to assess.

Box 10.5 gives an example for the necessary steps undertaken in a DCF analysis.

BOX 10.5   PERFORMING A DCF ANALYSIS OF AN ALUMINIUM SMELTER IN ICELAND

Iceland is a country in possession of rich supplies of energy. Renewable energy sources, consisting of geothermal and hydro power, provide almost all of Iceland's electricity. Imagine the government evaluating a join venture investment project in an aluminium smelter. Aluminium smelters are heavy industrial plants consuming large amounts of energy in order to produce aluminium. In addition, they are very large investment projects, with capital expenditure (CAPEX) often ranging beyond 1 billion US$. To calculate the financial return of such a project, the government and its joint venture partner could perform a DCF analysis pursuing the following steps:

1. First, the investors have to make an assessment of the initial CAPEX of the project. This will include the costs for purchasing machinery, constructing the plant and providing adequate infrastructure in the form of roads, rail or even a port.
2. In addition, the government and its partner enterprise will have to evaluate future revenues of such an investment. These will be driven by the future price of aluminium and the expected aluminium volume which can be sold in the market. Trend-line analysis or modelling could be used to forecast both volumes and prices.
3. Further, the investing parties will have to calculate operating costs of the aluminium smelter.
4. On the basis of steps (1) to (3) and other assumptions regarding cash-relevant items such as taxes, free cash flows of the investment opportunity are calculated for each future year. Free cash flows have to be discounted, for example based on the weighted average cost of capital.

5. If the sum of all future discounted cash flows is positive and if it exceeds that of similar investment projects, the investment opportunity could be attractive. Nevertheless, the government should also bear in mind potential external effects such as the environmental damage of such an investment and try to quantify it. For example, aluminium smelters produce a quantity of fluoride waste, which, unless carefully controlled, can be very toxic to vegetation around the plants. In order to see if the results remain stable against changes in the assumptions, the government should additionally develop scenarios.

10.2.6 CROSS-IMPACT ANALYSIS

Table 10.1   Cross-impact matrix

	E1	E2	E3	E4
E1				
E2				
E3				
E4				

New technologies cause changes and these changes themselves may affect other parts of a system. Therefore one does not only need to know the direct impact of new technologies but also the impacts of the impacts (see Braun 1998: 128).

These effects can be shown by constructing a matrix which allows us to register cross-impacts (see Table 10.1). The intensity of influence of events or trends of the first column on events or trends in the respective row can be marked in each field. One can try to quantify the size of impacts via attributes such as 'weak' or 'strong' or via numbers on a scale of 1 to 5. It is also possible to allocate probabilities to the occurrence of events. However, excessively attributing numerical values might give a false sense of precision. Cross-impact analysis is useful because it forces managers to think about unexpected cross-impacts.

10.2.7 UTILITY ANALYSIS

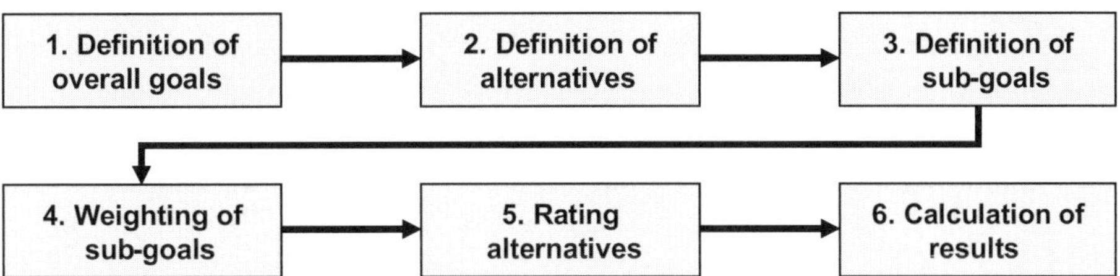

Figure 10.5   Steps of a utility analysis

Utility analysis is a practical tool supporting managers in decision-making processes. It can be used to make a selection between different technologies or between different sources of technology. Utility analysis divides the overall goal of a project into sub-goals (evaluation criteria). Managers asses the utility of each alternative for each sub-goal and then aggregate the scores to an overall utility score for each alternative. Utility analysis consists of the following steps (see Figure 10.5):

- Step 1: definition of overall goal of the project.

- Step 2: definition of alternative technologies or sources of technology.

- Step 3: definition of sub-goals (evaluation criteria).

- Step 4: weighting of sub-goals.

- Step 5: rating alternatives with regard to each sub-goal.

- Step 6: calculation of overall score for alternatives.

In the following, we will illustrate each step with the help of a fictive case.

10.2.7.1 Step 1: Definition of the Overall Goal

In the first step of utility analysis, the overall goal of the project needs to be defined.

Imagine ArgEnergy is a local Argentinean company wanting to acquire wind turbine technology. Wind turbine technology is concerned with the extraction of energy from the wind. All wind turbines installed by the end of 2008 worldwide are generating 260 TWh per annum, equalling more than 1.5 per cent of the global electricity consumption (WWEA 2009: 4).

As costs for wind energy installations have decreased dramatically and as wind technology can substantially reduce the emission of greenhouse gases, ArgEnergy estimates a great market potential for wind turbine technology in Argentina. The goal of the company is to gain possession of some form of wind turbine technology.

10.2.7.2 Step 2: Definition of Sourcing Alternatives

In a second step, different decision-making alternatives have to be defined. The management of ArgEnergy, together with the local investment promotion agency, discusses three different alternative sources for wind turbine technology:

- The first alternative is the negotiation of a *licensing agreement* with an American manufacturer of wind turbines. A license includes the use of patents, training of personal, and the provision of different forms of explicit technology (handbooks, software tools, etc.) necessary for manufacturing and operating the wind turbines.

- The second option consists of the *establishment of a joint venture* with a German producer to undertake common economic activity in Argentina. A new legal entity would need to be established to which both parties contribute equity and from which they share profits and risks.

- The third option is the attraction of a *wholly owned subsidiary* of a Danish wind energy manufacturer to Argentina. ArgEnergy could function as a main supplier for this firm, trying to gain insights into the turbine technology through vertical spillover effects (see Section 6.3).

10.2.7.3 Step 3: Definition of Sub Goals (Evaluation Criteria)

Now the overall goal of the investment project needs to be broken down into sub-goals. The following should be considered:

- criteria need to be logically deducted from the overall goal of the project

- criteria need to be independent of each other

- the number of criteria should not be too high as this will lead to great complexity

In our case, the management of ArgEnergy defines the following criteria:

- acquisition of *assembling technology* for wind turbines

- acquisition of *manufacturing technology* for turbine key components

- acquisition of *R&D capabilities* for wind turbines

- minimizing investments for ArgEnergy

- gain long-term *strategic independence* from other companies

10.2.7.4 Step 4: Weighting of Sub-goals

In a fourth step, evaluation criteria have to be weighted. The summarized utility of all criteria is 100 per cent, therefore percentage weights have to be distributed among the different criteria.

The management of ArgEnergy performs the weighting of criteria in a workshop. Table 10.2 shows the results of the weighting. Reasons for the weighting are given in the right column.

Table 10.2   Weighting of evaluation criteria

Criteria	Weighting (%)	Reasons for weighting
Acquisition of assembling technology	5	ArgEnery is not so much interested in the acquisition of assembling technology because of low profit margins in this business.
Acquisition of manufacturing technology for key components	30	ArgEnergy has some experience in the plastics and aluminium industry. Some key components of wind turbines such as blades are made of glass-reinforced plastic and aluminium. Due to this prior technology base, the company sees good chances for the absorption of manufacturing technology for components. In addition, profit margins in this business are high.
Acquisition of R&D capabilities	15	ArgEnergy aims at a long-term involvement in the field of wind energy. The management is of the opinion that this can only be achieved with the acquisition of own R&D capabilities.
Minimize investments	20	The company has made some substantial investments in other fields in the past. The available financial options for future investments are limited.
Gain strategic independence	30	The management gives great importance to strategic independence of ArgEnergy because of management culture and historical experience.

10.2.7.5 Step 5: Rating alternatives with regard to each sub-goal

Now the different alternatives have to be rated. This can be done by applying points on a scale from 1 (very weak) to 5 (very strong). Ratings have to be performed for each criterion.

In a workshop, the management of ArgEnergy discusses the ratings for the alternatives. Finally, the management agrees on the results presented in Table 10.3. For example, managers think that the licensing agreement offers few possibilities to develop their own R&D capabilities and to become strategically independent (rating 1), because ArgEnergy would always depend on the technology transfer of the American provider and its training. Managers also see licensing as very investment intensive (rating 1) while they view the setup of a supplier division as less costly (rating 3), due to already existing production facilities in the field of aluminium and plastics.

Table 10.3   Utility rating of alternatives

Criteria	Licensing agreement	Joint venture	Supplier for wholly owned subsidiary
Acquisition of assembling technology	3	4	1
Acquisition of manufacturing technology for key components	5	4	5
Acquisition of R&D capabilities	1	4	3
Minimize investment	1	2	3
Gain strategic independence	1	2	4

10.2.7.6 Step 6: Calculation of overall rate for alternatives

After having performed the rating of alternatives, overall results have to be calculated. The following calculations have to be made:

- the rating for each alternative has to be multiplied with the weight of criteria

- all weighted ratings for an alternative need to be summarized

- the alternative with the highest number of points is the preferable one

The results for ArgEnergy are provided in Table 10.4. As can be seen, the option to work as a supplier for a wholly owned subsidiary provides the highest utility score and is preferable. The CEO of ArgEnergy decides to realize this alternative.

Table 10.4   Overall results of utility analysis

Criteria		Licensing agreement		Joint venture		Supplier for wholly owned subsidiary	
Description	**Weight (%)**	**Simple rate**	**Weighted rate**	**Simple rate**	**Weighted rate**	**Simple rate**	**Weighted rate**
Acquisition of assembling technology	5	3	15	4	20	1	5
Acquisition of manufacturing technology for key components	30	5	150	4	120	5	150
Acquisition of R&D capabilities	15	1	15	4	60	3	45
Minimize investment	20	1	20	2	40	3	60
Gain strategic independence	30	1	30	2	60	4	120
Overall points			**230**		**300**		**380**

In summary, utility analysis may provide a valuable tool to structure complex decisions such as the selection of a technology or the source of a technology. It is capable of performing a quantitative analysis with non-monetary values, systemizes the decision to be taken, and allows a direct comparison among criteria.

However, it runs the risks of relying too much on quantitative results instead of common sense, it is time- and labour-intensive, and the results depend to a large degree on the criteria weighting applied. To level the last problem, you may perform a scenario analysis and see what happens if the weight of criteria is altered.

10.3 Best Practices of Technology Assessment

The methods presented above allow managers to structure the process of technology evaluation. In addition, the following best practices can support you in assessing technology (OECD 1998):

- Create an 'evaluation culture' by stimulating awareness, positive attitude, a knowledge pool and expertise to conduct evaluations.

- Formulate guidelines and establish consensus on approach and measures to secure acceptance of results.

- Establish multidisciplinary evaluation teams to be able to judge on scientific, economic, managerial, and political dimensions of technology transfer.

- Secure the independence of evaluators and avoid allowing particular interests to dominate the process.

- Carry out evaluations regularly to achieve economies of scale for the exercise and to accumulate knowledge.

- Secure the provision of adequate data to allow the application of more advanced quantitative methods.

11

Technology Attraction

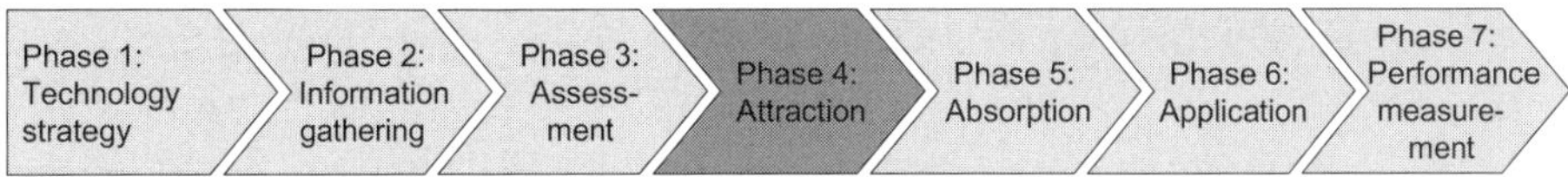

Figure 11.1   Technology attraction (phase 4)

In phase 3 we saw how a technology can be assessed and a decision for a particular technology can be taken. This chapter gives an overview of the attraction of technology (phase 4 in Figure 11.1). The attraction of technology refers to the direct, internal transfer of technology from a MNC to its subsidiary (Figure 1.2, see also Section 6.1).

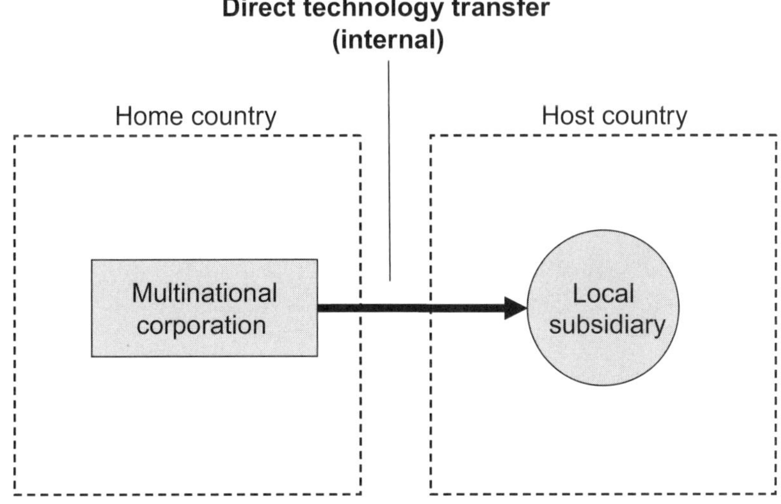

Figure 11.2   Internal technology transfer

In order for such a direct transfer to take place, a MNC needs to take the following two steps:

- First, a MNC needs to make a direct investment into the host economy by setting up a local subsidiary.

- Secondly, a MNC has to decide to transfer technology (in the form of machinery, equipment, trainings, processes, skilled personal, etc.) to this subsidiary.

Governments can try to trigger these two steps by the following policies:

- First, governments may try to improve their overall attractiveness for FDI in order to move a MNC to set up a local subsidiary. Issues to be addressed are the general attitude towards FDI, the establishment of an investment promotion agency, transparency, property rights, etc. We will discuss the different measures that can be put forward in Section 11.1.

- Secondly, governments can aim to directly set incentives for the technology transfer from MNCs to local subsidiaries. Such incentives can be in the form of tax reductions or transfer requirements. We will discuss the various available measures in Section 11.2.

As negotiations play an important role in attracting FDI and technology, we conclude this chapter by providing some methods and concepts in this field (Section 11.3).

11.1 Policy Measures for Attracting FDI

We have already shown in Section 5.4 that FDI depends on the expected financial return which is influenced by the following factors:

- macroeconomic conditions;

- attractiveness of the local market;

- factor conditions; and

- investment and trade policies.

Not all of these factors are under the direct influence of government bodies and investment promotion agencies. For example, the interest rate of a country is usually set by an independent central bank. Equally, a country cannot directly influence its geographic location or its population size. Nevertheless, there exist a variety of factors which governments and investment promotion agencies can influence. These fall in the field of economic policy as well as investment and trade policy (see UNCTAD 2003b, Moran et al. 2005: 378):

- *General attitude towards FDI.* The general attitude of a government towards foreign investors is of major importance in attracting investments to a country. Measures such as the establishment of international conferences, documents on the country's attractiveness, or statements by the highest authorities provide indicators for foreign investors that the government is committed to FDI.

- *Setting up of an investment promotion agency.* Investment promotion agencies communicate and market existing investment opportunities. They may also draw the attention of relevant government bodies to areas that are important for making a location more attractive. Thus investment promotion agencies can serve as a bridge between the private and public sector.

- *Corruption.* A high level of corruption is often one of the major obstacles for FDI. Typical problems of corruption are threats of over-taxation unless an informal payment is made, bribes required for establishing a business, or contracts not decided on the basis of price. A widely used measure of corruption is the Corruption Perception Index calculated by Transparency International, a non-profit organization. The index shows how countries compare against each other regarding the perceived level of public-sector corruption.

- *Transparency of laws and regulations.* Heavy bureaucracy and the lack of transparency have been serious deterrents for FDI. Investors perceiving investment conditions as unclear or bureaucracy as unreasonably high might not invest in a country.

BOX 11.1   CHINA – SUCCESSFUL POLICY MEASURES VIA SPECIAL ECONOMIC ZONES

In the mid-1980s, China introduced what was to become a highly successful model to attract FDI, the 'special economic zones'. Originally, these zones were meant to be experiments with economic reform, at the time when Chinese economic policy was still quite Communist. Over time, the zones became havens for foreign investors. Foreign investors received special treatment, including insulation from the complicated regulations associated with Communist doctrine. The zones also provided an upgraded infrastructure, tax relief, duty-free importation of capital goods, provision of trained workers and insulation from corruption. As a result, the 'special economic zones' attracted a cumulated US\$ 6 billion in FDI, enormous figures compared by the standards of the 1980s.

Moran et al. (2005: 378).

- *Protection of property rights*. MNCs want to be sure that the assets they are buying locally, or transferring from abroad, remain in their possession and under full control. This requires an independent and efficient judiciary system, protection against criminality, solid and transparent ownership laws and an effective system for registering ownership.

- *National treatment*. National treatment for MNCs means that there are no special performance requirements, all sectors are open for investments and no requirements to purchase from local sources exist.

- *Conversion and transfer policies*. The ease of conversion of foreign into local currency and the possibility of transferring funds in and out of a country are necessary conditions for investors. In many developing countries, FDI and technology transfer have suffered serious setbacks because of inappropriate exchange rate policies and allocations (Cohen 2004: 195).

- *Dispute settlement*. Typical themes of business disputes might be non-payment of debts or non-delivery of goods and services. As disputes may not always be settled informally, it is in the interest of every investor to ensure that the legal system for settling disputes is both well functioning and fair. Box 11.2 elucidates dispute settlement in Malaysia.

BOX 11.2   DISPUTE SETTLEMENT IN MALAYSIA

Malaysia is signatory to the UN-sponsored convention on the settlement of investment disputes. The domestic legal system is open and accessible. Past cases of foreign investment dispute, which have been rare, have consistently been handled satisfactorily by existing dispute settlement mechanisms. Should local administration and judicial facilities fail to satisfy claimants, a dispute would be submitted to the International Center for Settlement of Investment Disputes (ICSID) under the aegis of the United Nations.

Adapted from UNCTAD (2003b: 13/7).

11.2 Policy Measures to Stimulate Technology Transfer via FDI

Once a MNC has decided to invest into a local subsidiary in a host country, a government is interested in encouraging a MNC to transfer technology to its subsidiaries. In Section 6.1 we discussed a variety of factors influencing MNCs decisions to transfer technology. Some of these, such as the global strategy of a MNC, are beyond the control of the government of a host country. However, there are three key factors which can be directly influenced by government bodies:

1. governments can influence the benefits of technology transfer via setting transfer *incentives* (see Section 11.2.1);

2. governments may establish certain *requirements* for technology transfer (see Section 11.2.2); and

3. governments can reduce the risk of technology transfer and theft by establishing and enforcing clear *intellectual property rights* (see Section 11.2.3).

11.2.1 TECHNOLOGY TRANSFER INCENTIVES

Technology transfer incentives aim directly at attracting technology related FDI by providing either financial or fiscal incentives. Most developed countries and a growing number of developing countries make use of these (see UNCTAD 2005a: 217):

- *Financial incentives* refer to direct funding of R&D projects by the government through the granting of preferential loans or subsidies.

- *Fiscal incentives* are tax-based. They can consist of accelerated depreciation (faster depreciation rates for R&D expenditures), tax allowances and credits (deduction of R&D expenditures from taxable income), tax holidays (exemption of firms investing in R&D from paying taxes for a given period of time), income tax allowances (targeted at personnel and products linked to the R&D activities of the firms), and import tariff exemption (for R&D-related imports).

However, empirical evidence suggests that financial and fiscal incentives only have a marginal impact on technology transfer (UNCTAD 2005a: 216). In addition, they impose the following risks (see Moran 2005: 308, UNCTAD 2005a: 216):

- international competition among countries in offering incentives could result in the wasting of public funds;

- firms may try to label costs not related to technology transfer as expenditures in order to benefit from favourable tax treatment; and

- there is a risk that a government might end up supporting R&D projects that firms would have undertaken even without support.

BOX 11.3   R&D INCENTIVES IN ASIAN COUNTRIES

Many Asian countries have adopted incentives to promote R&D:

- In China, MNCs can set up R&D centres as independent entities (under the rules applying to Sino–foreign joint ventures), wholly foreign owned enterprises or as independent departments or branches of existing companies. Equipment and parts imported by R&D centers meeting certain requirements are exempt from customs duties and import value added tax and the technologies they develop and use are exempt from business tax.
- India offers a ten-year tax holiday for companies engaged exclusively in scientific R&D with commercial applications.
- In Malaysia, companies can offset 100 per cent of capital expenditure incurred within ten years against 70 per cent of their income.
- Singapore allows a 100 per cent deduction of R&D expenses (in certain cases 200 per cent) and provides various grants and tax exemptions.
- Thailand revamped its system of R&D incentives in 2004, after which firms can be entitled to a corporate income tax holiday for up to eight years.

Adapted from UNCTAD (2005a: 217).

11.2.2 TECHNOLOGY TRANSFER REQUIREMENTS

In addition to setting incentives, countries may also develop requirements for foreign companies to undertake technology transfer to their economies. These usually take the form of R&D requirements, as in the following countries (see UNCTAD 2005a: 215):

- In India, R&D requirements have been imposed on both foreign and local investors to encourage the set up of in-house R&D facilities or the entrance into long-term consultancy agreements with local R&D institutions. However, requirements have tended to be minimal and are seldom closely monitored.

- In China, requirements to undertake R&D are imposed as a condition for entry in selected industries where the inflows of FDI may be considerable but where MNCs have not undertaken R&D activities. A prominent example is the passenger car industry. In an effort to tackle the relatively slow enhancement of domestic innovation capability, the 2004 industrial policy required the establishment of an R&D centre with an investment of at least 500 million yuan (about US$ 60 million) for any new automotive project to be approved.

However, the use of performance requirements brings the following disadvantages:

- Requirements are difficult to implement and control, as technology transfer is hard to measure and quantify.

- Requirements always risks driving back some FDI. While this risk might be lower for countries with a stronger bargaining power vis-à-vis the foreign investors (because of a large domestic market such in the case of China and India), the use of mandatory requirements in smaller economies increases the risk of losing FDI (UNCTAD 2005a: 215).

- Empirical evidence on the effectiveness of technology transfer requirements is not encouraging. A variety of studies shows either no positive or even a negative impact of performance requirements on technology transfer to host countries (e.g. Urata and Kawai 2000, Blomström et al. 2000: 216–217, see Moran 2005: 293 for an overview).

11.2.3 INTELLECTUAL PROPERTY RIGHTS

A well-defined, balanced and enforceable intellectual property rights (IPR) regime influences a MNC's decision to transfer technology by lowering the risks of technological theft. Nevertheless, there is some debate on the value of intellectual property rights for less developed countries. Evidence suggests that strong IPR only begin to benefit countries with per capita incomes above US$ 7,750, as they move away from building local capabilities through copying and begin to engage in more innovative activities (Smith 2009: 89):

- On the one hand, in order to build up innovative capabilities and formal R&D, an enforceable system of IPRs is indispensable (Hoekman et al. 2004: 14). Without sufficient protection of technologies, companies might actively hold back their technologies and impede technology transfer (see Box 11.5). Empirical studies show that an increased effectiveness of intellectual property rights in 16 countries has resulted in MNCs transferring more technologies to reforming countries (Branstetter et al. 2006).

- On the other hand, countries with loose IPR might acquire technological capabilities via reverse engineering (as shown in Box 2.3). In addition, strong IPRs might at the same time place burdens on consumers by assigning the owner of intellectual property a degree of monopoly power. However, loose IPRs set strong incentives for MNCs to hold back technology, as shown in Box 11.5.

Regardless of these arguments, the agreement on Trade-related Aspects of Intellectual Property Rights (TRIPS) requires all members of the WTO to meet minimum standards of IPR protection. The main areas of intellectual property include copyright, geographical indications, patents, trademarks and undisclosed information (including trade secrets).

BOX 11.4   THE REPUBLIC OF KOREA'S (SOUTH KOREA) CHANGE IN NATIONAL POLICIES

Korea is a country that successfully moved up the technology ladder, also due to its capability to adopt policies. In the beginning the intellectual property rights of the country were weak. Existing policy encouraged adoption and imitation of technologies foreign firms had permitted to enter the public domain or were willing to provide cheaply.

As production processes matured, the government promoted the development of technical skills through education and workplace training and ensured the absence of anti-export bias.

In the 1980s, Korea shifted to creative imitation, involving more significant transformation of imported technologies. This required domestic R&D and in-house research capabilities to adapt technology. Government polices during this stage changed by welcoming more formal channels of technology transfer and strengthening the intellectual property rights regime.

Adapted from Hoekman et al. (2004: 18).

BOX 11.5   THE FEDERATION OF GERMAN INDUSTRIES ON TECHNOLOGY HOLD-BACK

The Federation of German Industries (BDI) is probably the most important and influential industry associations in Germany. Due to a German chancellor's visit to China in 2006, the association was actively lobbing against 'enforced' and 'illegal' technology transfer in China, as a BDI representative called it. The BDI suggests to German firms to protect their technologies through a variety of measures:

- hold back key technologies
- only transfer technologies for a limited time and with restricted licenses
- divide a product into several modules. This way, suppliers only gain access to part of the product technology
- exclusively integrate and test product modules in the home country
- only provide suppliers with documentation, strategies, client relations, and know-how if absolutely necessary
- do not send technical documents without protection
- train engineers on risks of technology theft
- do not accept every contract
- do not sanction your distributors if they decline a sales opportunity for reasons of technology protection

Adapted from FAZ (2006: 13).

11.3 Negotiations

The Sections above focused on policy issues of technology attraction for which investment promotion agencies can use their policy advocator role. This Section deals with the negotiation of transfer projects. Governmental organizations and investment promotion agencies may support local firms in technology attraction by assisting them in negotiations with foreign partners. They may also conduct negotiations themselves when trying to attract foreign investments to their country. Some of the issues that require negotiation might be: the incentives granted to a foreign investor, the type, amount and quality of a technology transferred, local firms' access to this technology, the money actually invested, or the number of jobs created.

11.3.1 PREPARATION OF NEGOTIATIONS

Negotiations should be planned carefully before they start by addressing the following issues:

- *Know and improve your BATNA*: BATNA is an acronym for 'best alternative to a negotiated agreement' (see Fisher et al. 2004). It is the preferred action in case a deal is not concluded in the negotiations. If the BATNA is strong, one can negotiate for more favourable terms, knowing that one has something better to fall back on in case a deal cannot be agreed (Luecke 2003: 17). If the BATNA is weak, managers should search for further alternatives to the deal. Even if only one alternative is available it will substantially improve the negotiation strategy.

- *Define the 'walk-away price'*: the 'walk-away price' is the least favourable point at which one is willing to accept a deal. This price should be derived from the BATNA and has to be determined before entering negotiations.

- *Understanding the other party*: it is essential that you study the interests and culture of the other negotiating party. Some key questions to be answered are: what is the other party's BATNA? What negotiation style does it follow? What are its financial conditions? Who decides what? (Spangle and Isenhart 2003: 72).

- *Understand the other party's culture*: cultural differences can influence technology transfer negotiations in significant and unexpected ways. For example, take an American salesman presenting a multi-million-dollar proposal to a potential Saudi Arabian client in a pigskin binder, considered vile in many Muslim cultures. He was unceremoniously tossed out and his company blacklisted from working with Saudi businesses. It is therefore important to gather information on key characteristics of the other party's culture, for example the degree of formality, greetings, the body language or emotions (Sebenius 2002: 76–80).

- *Preliminary proposal*: before beginning negotiations on a transfer project, each party needs to determine its objective for concluding a deal. This objective is reflected in a preliminary proposal submitted to the other party, expressing interest in acquiring certain rights, know-how, or skills (UNIDO 1996).

- *Definition of planning and negotiation team*: the planning team should consist of the chief negotiator, a technical expert, a financial expert and a legal expert. The planning team has to analyse all parameters for a proposed agreement so the negotiating team has the information required to present its position to the other party. The negotiation team should be kept as small as possible and has to speak with one voice (UNIDO 1996).

11.3.2 NEGOTIATION TECHNIQUES

During the negotiation process, a variety of techniques can be used to resolve issues and conflicts fairly and to advance negotiations on technology transfer:

- *Making a good start*. Once a party has decided to negotiate, it is important to get a good start. This can be done by expressing respect for the other party's experience and expertise, addressing the content of the negotiations positively, framing it as a joint endeavour and by emphasizing the openness to the other side's interests and concerns (Luecker 2003: 48).

- *Anchoring the negotiation*. Anchoring is an attempt to establish the negotiations around a reference point. In some negotiations, you might gain an advantage by putting the first offer on the table because this offer can become a psychosocial anchor. Studies show that negotiation outcomes often correlate to the first offer (Luecker 2003: 50).

- *Defer difficult issues*. Probably the most useful technique for advancing the process of reaching agreement is to defer those issues that appear most difficult to resolve and tackle those that can be settled quickly. Experience has shown that a series of agreements on lesser issues creates a momentum that induces negotiators to reach an agreement on the difficult issues. This should be considered when setting the agenda (UNIDO 1996).

- *Keep track of concessions*. In negotiations a manager will likely have to make concessions. However, they should keep a summary of the concessions made. This serves as self-control, proves willingness to compromise and may help obtain concessions from the other party later in the negotiations.

- *Spread the concession out.* Sometimes negotiations require a concession by the other party that is so large there is little chance of obtaining it. A technique that can be used in such a case is to break down the issue into various components and then to spread concessions on the relatively minor components through the various negotiating sessions (UNIDO 1996).

- *Building long-term commitment.* Technology transfer usually involves a long-term relationship between parties. Managers should keep this in mind during negotiations. Managers should therefore dedicate some thought to the mechanisms that should be in place during the term of the agreement. These guarantee that the counterpart is fulfilling their part of the agreement (Cohen 2002: 162).

12

Technology Absorption

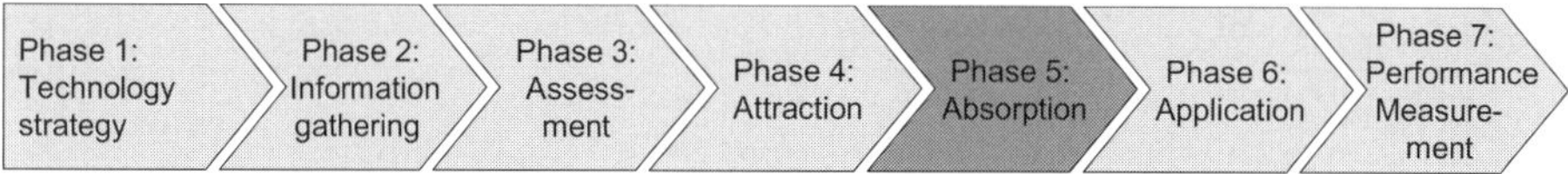

Figure 12.1   Technology absorption (phase 5)

In Chapter 11 we laid out a variety of policy measures that may be applied to support the attraction of technologies (phase 4 in Figure 12.1). These aim to engage MNCs in an *internal transfer* of technology from the parent to the local subsidiary (see also Section 6.1). However, an internal transfer to a subsidiary does not yet guarantee a successful absorption and diffusion of a technology within the economy of a host country (phase 5 in Figure 12.1). For this to occur, external spillovers have to materialize. Without external spillovers, the technology transfer is restricted to the MNC, only generating limited benefits for the host country. In the extreme case of a MNC shutting down its local subsidiary, the technology would even be lost for the host country.

In this chapter we therefore focus on policies that may support the absorption of technology by the host country through horizontal (to competitors, see Section 6.2) or vertical spillovers (to suppliers or customers, see Section 6.3). Two policy dimensions influence this process:

- Countries may build their *absorptive capacity* by investing in their human resources and public R&D capacity as well as by enhancing private R&D (Section 12.1).

- Countries may generate opportunities for spillovers by providing *fields of interaction* between MNCs and local organizations through the establishment of science parks and industry clusters (see Section 12.2).

12.1 Building Absorption Capacity

Absorptive capacity refers to the ability to recognize the value of a new technology and to assimilate it (see Chapter 4). There is wide empirical support that absorptive capacity of host countries is crucial for obtaining significant benefits from technology transfer in FDI (see Saggi 2000 for an overview). Countries will only benefit from technology spillovers if they possess the required capacity to assimilate the respective technology.

A country can influence its absorptive capacity with the following policies:

- via the development of human resources (Section 12.1.1);

- through the investment in public R&D capacity (Section 12.1.2); and

- by encouraging private R&D (Section 12.1.3).

12.1.1 DEVELOPMENT OF HUMAN RESOURCES

Human resources are of critical importance to build absorptive capacity for technology transfer (see for example Saggi 2000). A country's improved supply of skilled people may occur as a result of long-term policies to raise educational standards on the one hand (see Section 12.1.1.1), as well as from efforts to attract human resources from abroad on the other hand (see Section 12.1.1.2).

12.1.1.1 Educational Standards

Education and training constitute perhaps the single greatest long-term leverage point available to governments in upgrading industry (Porter 1990: 628). Educational needs differ between distant stages of economic development and technological specializations. Policy-makers have to ensure that the education system delivers the kind of skills that are most in demand, requiring a match between the technology strategy and the educational policies (see Chapter 8). Box 12.1 shows how this is done in Singapore.

BOX 12.1   MONITORING FUTURE SKILL NEEDS IN SINGAPORE

In Singapore the Ministry of Trade and Industry, the Economic Development Board and the Council for Professional and Technical Education work closely together to monitor future skill needs. The institutions draw on inputs from foreign and local investors as well as from education and training institutions. This information is matched against national policy objectives and used to build targets for various universities, schools, and the Institute for Technical Education. As a result, the number of students enrolled in universities has increased from 42,971 in 1998 to 69,028 in 2008, a total increase of approximately 61 per cent in only 10 years.

UNCTAD (2005a: 203), Singapore Department of Statistics (2009).

The effectiveness of an educational system is partly a function of the rate of spending, but also of sound educational policies. Some of these are (Porter 1990: 628–630, UNCTAD 2005a: 204):

- *Educational standards are high.* The educational and training systems must demand high performance and students must have to compete for advancement. For example, in Germany the ministry of education enforces certain standards for the final examinations of high school students.

- *Teaching is a prestigious and valued profession.* A high-quality education requires a cadre of well-prepared and competent teachers at all levels. For example, in Japan and Korea, teaching at all levels is prestigious, and faculity positions are staffed with outstanding people.

- *The majority of students receive education and training with some practical orientation.* The majority of students must be given a foundation that will allow them to participate actively in the economy. Maths, computing, writing, basic sciences and languages are of particular importance.

- *Respected and high-quality forms of higher education besides the university exist.* Technology transfer does not require that all students persue an advanced degree in universities. Technical universities and vocational schools are equally important as they provide the skills for continued personal development.

- *There is a close link between educational institutions and the private sector.* A close connection between educational institutions and private companies guarantees the relevance of the skills provided by these institutions. An example is the German apprenticeship system combining education with on-the-job training over a period of three years.

- *Educational policies are regularly updated.* The regular and continued update of skills is important, especially when there is mismatch between the supply and demand of specialized skills. Education policies therefore need to evolve over time as the demands from industry change and countries develop. Box 12.2 shows how Korea has made various policy shifts due to its economic development.

BOX 12.2   KOREA'S POLICY SHIFTS IN EDUCATION

The Republic of Korea's has modified its education policies several times to reflect new requirements:

- In the 1960s, a system of technical training was set up as part of broader efforts to improve the infrastructure for science and technology.
- In the 1970s, the government emphasized technical and engineering education in the fields of heavy and chemical industries.
- In the 1980s, focus shifted towards technology-intensive industries and greater efforts were made to bring back Korean scientists from overseas.
- Since 1990, more attention has been given to promote creativity, for example by formulating the Creative Research Initiative in 1997, encouraging a move from 'imitation' to 'innovation'.
- More recently, special incentives have been offered for universities to become less teaching- and more research-oriented.

Adapted from Vertzberger (2005: 24–25).

Policy intervention may be needed to re-skill and retrain production workers, technicians and engineers. As Box 12.3 indicates, countries can involve foreign affiliates in this process, for example by encouraging them to participate in joint projects with universities and other training institutions.

BOX 12.3   EXAMPLES FOR JOINT PROJECTS BETWEEN FOREIGN AFFILIATES AND THE EDUCATION SYSTEM

Some countries have achieved close collaboration between public institutes and foreign affiliates to upgrade employees' skills:

- Costa Rica attracted a major semiconductor investment from Intel in 1996. Close links between Intel and the Instituto Tecnológico de Costa Rica helped secure financial support from Intel to develop new programmes and increase enrolments of engineering students.
- The auto parts maker Delphi collaborated with the privately run Tec de Monterrey in Mexico to ensure adequate skills for its development work in Ciudad Juarez.
- In India, Motorola worked with the Pune Institute of Advanced Technologies to offer a postgraduate degree in advanced telecommunications engineering with a software focus.
- In Singapore, the efforts of the Economic Development Board to involve MNCs and foreign governments in training programmes helped ensure that they were relevant and up to date.

Adapted from UNCTAD (2005a: 203).

12.1.1.2 Attracting Human Resources

In many cases countries cannot create all the skills they need by themselves. Many countries have therefore made the attraction of talent a key priority. For instance, Singapore has a liberal immigration policy to attract highly skilled people to private firms and public research institutes (UNTCAD 2005a: 205). By 2003, close to a third of doctorate-level research and engineering scientists in the tertiary and public research institutions in Singapore were non-citizens. This migration has led to Singapore having the seventh highest ratio of researchers in the world (UNCTAD 2005a: 205).

Increased mobility of highly skilled workers has a negative and a positive effect for developing countries (see UNCTAD 2005a: 206):

- Mobility may cause a 'brain drain' from developing economies, aggravating an already limited supply of skilled human resources.

- Skilled workers from developing countries may return to their home country, with potentially new skills, entrepreneurship, knowledge and capital. For example, Bangalore in India has some 35,000 'returned non-resident Indians'. Many of these have training and work experience in the United States. If countries can create conditions that are conducive to such return flows, the original brain drain can be turned into brain circulation with positive implications for technology transfer.

Box 12.4 explains the policies of the Republic of Korea to attract back scientists.

BOX 12.4   POLICIES IN THE REPUBLIC OF KOREA TO ATTRACT BACK SCIENTISTS

In the 1960s the Republic of Korea initiated a project to recruit Korean scientists working abroad. These efforts began with the establishment of the Korea Institute of Science and Technology in 1966, and in 1968 a specific project was launched to attract back qualified scientists. As inducement measures they were offered modern laboratories, competitive salaries and autonomy in their research. From 1968 to 1979, 238 scientists returned to stay permanently in the country and another 255 scientists returned temporarily. These researchers played an important role in the 1970s and 1980s and contributed to cultivating new human resources in R&D.

Adapted from UNCTAD (2005a: 206).

12.1.2 INVESTING IN PUBLIC R&D CAPACITY

In addition to setting educational standards, countries must also raise their absorptive capacity by investing in public R&D capacity (Hoekman et al.

2004: 16). Public research institutes and universities can perform the following functions regarding technology transfer (see Porter 1990: 632, Feldman and Kogler 2008: 446):

- Undertake basic research producing new technology, increasing the stock of knowledge.

- Provide technical services, for example testing and consultancy, for firms.

- Provide training for graduates.

- Generate new instruments and methodologies that may be applied in the private sector.

- Create new firms, for example as spin-offs from university research projects.

Foreign affiliates of MNCs can interact with public research institutes and universities by subcontracting services to them, by undertaking joint research projects or programmes, and by employing skilled people from the institutes. India gives a good example of how the relevance of the public research institutes for the private sector has been improved (see Box 12.5).

BOX 12.5   PUBLIC RESEARCH INSTITUTES IN INDIA

India has a network of 38 laboratories and 45 field/extension centers under the Council of Scientific and Industrial Research (CSIR), employing over 4,600 active scientists. Because the system had until then produced little technological benefit to industry, the government launched a major reform programme the late 1980s. It limited the level of public financing of the laboratories, and set a target for CSIR to earn 40 per cent of its expenditures by selling research and other services to industry. The new annual budget of each laboratory was determined by its revenue-earning capability. As a result, the institutes' earnings almost tripled between 1992 and 1997. By 2005, CSIR accounted for around 25 per cent of all patents filed in India by locals.

Adapted from UNCTAD (2005a: 207–208).

12.1.3 ENCOURAGING PRIVATE R&D

The third trigger to increase absorptive capacity is the encouragement of private R&D, for example via incentives. Such incentives can take the form of *direct research grants, subsidies, or tax credits* to domestic firms (Porter 1990: 634). However, subsidies contain the problem that the commercial prospect of research projects is difficult to evaluate. Without the financial risk, firms might propose bad projects, not manage projects well, or use funds for projects they would have conducted anyway (Porter 1990: 634). The most effective way to encourage private R&D seems to be partial funding of specialized research institutes connected to industry clusters, which we will describe further in Section 12.2.1.

12.2 Facilitating Interaction

Theories of technology transfer stress the importance of interaction for technology absorption (see Chapter 4). Therefore, in order for technology spillovers to occur, a MNCs subsidiary must be closely integrated into the host country's economy (Hamar and Stephan 2006: 145). To facilitate such an integration, governments need to create platforms allowing close interaction on a vertical (to suppliers and customers) and on a horizontal level (to competitors). Two institutions may provide such platforms: industry clusters and joint ventures.

12.2.3 PROMOTING INDUSTRY CLUSTERS

A cluster is a geographic concentration of interconnected companies in a particular field. A cluster can include companies, suppliers, trade associations, financial institutions and universities in a field or industry (Dorf and Byers 2008: 167). Clusters are often built on one or two large firms, for example subsidiaries of MNCs, acting as leaders or anchor enterprises feeding the growth of numerous smaller enterprises. Geographic proximity creates opportunities for horizontal and vertical technology spillovers. In addition, clusters often provide a critical mass of talent, technology and suppliers in order to enter an industry (Dorf and Byers 2008: 167). A famous example of an industry cluster is the area south of San Francisco called Silicon Valley, a cluster for electronics and computer companies.

Several actions may stimulate the emergence of clusters (see Porter 1990: 632, Ferreira et al. 2006: 101, Giroud 2006: 188):

- Policy-makers can analyse the industrial landscape and *identify potential flagship enterprises* with 'fertility potential'. For these, governments can set a certain array of incentives, such as establishing free trade zones, to attract such an enterprise to a certain location (see Section 11.1).

- Governments have to provide an *adequate infrastructure* for industry clusters. This involves the removing of administrative hurdles for land acquisition and the provision of roads, rail, electricity, telecommunication or ports, if required.

- Governments can *promote linkages with universities and research institutes*, for example via financing of joined projects. Research institutes, which are both privately and publically funded, create a natural focus for solving industry problems and producing research relevant for a certain industry. In many countries trade associations play an important role in funding and creating such specialized research institutes.

- Governments can seek to *promote science parks* (see Box 12.6). In an UNCTAD survey, managers mentioned science parks as the second most commonly used policy tool to absorb technology and R&D from FDI (UNCTAD 2005a: 215). Science parks offer attractive features as they facilitate networking, offer access to skilled people and provide the necessary infrastructure and administrative support (UNCTAD 2005a: 218). According to its website, The International Association of Science Parks (IASP) alone accounted for approximately 370 member science parks with 200,000 companies located in these parks (2009).

- Governments can promote *linkages between MNCs' subsidiaries and local suppliers* by a range of programmes, for example via information exchange, tax incentives or vendor development schemes.

- Governments may *facilitate the emergence of spin-offs* and start-ups around MNCs' subsidiaries by grating start-up grants, tax incentives, or informational and knowledge support, for example in the form of business planning.

It is noteworthy that industry clusters do not form night to day. Rather, they take a long time to emerge and require the gradual trust building of all cluster participants (Ferreira 2006: 102).

BOX 12.6   SCIENCE PARKS IN VARIOUS COUNTRIES

Science parks exist in a variety of countries in Asia and Africa:

- A well-known science park in Asia is the Hsinchu Science Park set up in 1980 in Taiwan. While it was originally established with a view to serving local companies, non-Taiwanese companies have also been attracted. In 2004, 52 out of 384 companies in the park were non-Taiwanese.
- In Singapore, the first science park was also established in 1980 and now hosts 300 local and foreign companies.
- The Zhongguancun Science Park (Beijing) is China's first and largest science park with more than 14,000 high-technology firms, including 1,600 foreign affiliates.
- The offshoring of software development to India has often benefited from the presence of dedicated technology parks for IT services. As of 2003 there were 39 such parks, accounting for 80 per cent of all India's software exports in 2002–03.
- In South Africa a new park — The Innovation Hub — is the first African science park that is accredited internationally. Its main objective is to attract a variety of enterprises active in ICT, electronics, life sciences and aerospace.

Adapted from UNCTAD (2005a: 219).

12.2.2 JOINT VENTURE REQUIREMENTS

Theoretically, joint ventures offer the best mechanism for effective technology absorption due to their potential for close cooperation between MNCs and local counterparts (Müller and Schnitzer 2003: 18, see also Sections 4.1 and 5.3). For example, in a survey on technology transfer in the Iranian vehicle, chemical and electronics industries, Moeini and Zawdie (1998) showed joint venture initiatives had been more successful than other technology transfer mechanisms. In practice, however, MNCs in many cases would rather prefer to make their investments via wholly owned subsidiaries and have little or no joint venture involvement. Reasons for this are (Stopford and Wells 1972):

- Wholly owned subsidiaries are perceived as easier to control, to operate and to manage by a head office.

- The implementation of company policies among branches is seen as more effective with wholly owned subsidiaries than in joint ventures where problems of coordination exist.

- Services and managerial skills provided by local counterparts are often perceived as unsatisfactory.

- MNCs try to avoid the sharing of technology with local partners.

As a reaction to MNCs' tendency to establish wholly owned subsidiaries instead of JVs, some countries have formulated *joint-venture requirements or joint venture incentives such as* tax breaks, repatriation of profits, duty exemptions and R&D tax credits. For example, in the defence industry of Malaysia, the government requires joint venture defence contracts and vendor development agreements to include technology transfer conditions such as training, design and prototype development (Malairaja and Zawdie 2004: 241). However, such regulations are countered by MNCs:

- Studies show that the technology employed in joint ventures tends to be three to ten years behind the cutting edge for the industry. Also, the amount of technical training provided to local managers and workers is often a fraction of that received in wholly owned affiliates (Morgan 2002).

- Most joint ventures operating in developing countries tend to concentrate on activities that are labour-intensive and involve skills at the low end of technological capabilities. For instance in the electronics sector, joint ventures are often focused on assembly operations involving relatively few design or product development skills (Malairaja and Zawdie 2004: 236, 245).

Even strong policy incentives for MNCs to establish joint ventures might only lead to the transfer of old technologies or technologies at the assembling stage. This is particularly the case when the main purpose of the parent company's joint venture is to exploit the local market.

Box 12.7 illustrates some experience with joint ventures in Malaysia.

BOX 12.7   EXPERIENCE FROM JOINT VENTURES IN MALAYSIA

Malaysia is one of the leading recipients of FDI in the developing world due to a number of favourable factors such as business environment, supply of infrastructure, a well-educated labour force, incentive schemes and political stability.

Foreign companies have mostly sought Malaysian firms as joint venture partners to expand the market for their products, undertake joint production of consumer and industrial goods, manufacture components and parts for their parent companies and bid jointly for construction projects and consultancy services.

Generally speaking, Malaysian firms have had limited success in translating joint ventures into widespread innovation and technological progress. This is partly due to weak bargaining positions of technology recipients, the failure of policy to address innovation issues and the fragmentation of the relevant institutions.

Some international joint venture operations have nonetheless helped in enhancing the technological capabilities of local partners. For example, the Japanse–Malaysian joint venture in the production of motor cars (Proton) has greatly increased the capability of the local partner HICOM and other local vendors in the design and manufacture of car components and parts (see also our case study in Section 15.5).

The government has recognized existing problems with technology absorption in joint ventures. It passed a new science and technology policy in 2002, to create a growing awareness of the importance of innovation.

Adapted from Malairaja and Zawdie (2004: 241–244).

12.2.3 COMPETITION POLICY

Technology spillovers may not only occur on a vertical level to suppliers and customers but also on a horizontal level between a MNC's subsidiary and local competitors. Competition policy can therefore play a role in generating opportunities for technology externalities. Competition provides a general incentive for firms – be they foreign or local – to innovate, for example by encouraging them to invest in R&D and other innovatory activities. For example, empirical work has shown a positive correlation between the level of competition and innovative output (Geroski 1994, Blundell et al. 1999). This is particularly evident in developing countries and transition economies, where firms that face greater pressure, especially from MNCs, are more innovative than firms that feel less pressure (Carlin et al. 2001). Recent work has shown that stricter competition laws and better enforcement of those laws have a positive impact on innovation in low- and middle-income countries (Clarke 2005).

For promoting greater benefits from FDI, some applications of competition policy are particularly relevant. These include standard setting and patent pools, merger control and policies to address restrictive business practices.

12.3 Case Study: Thailand's Electronics Industry

In this Section we provide you with a practical case study on attracting and absorbing foreign technology. It deals with Thailand's strongly export-oriented electronics industry. The industry's competitiveness is largely based on the attraction and absorption of foreign technology (the following case is based on UNCTAD 2005b).

12.3.1 CHARACTERISTICS OF THAILAND'S ELECTRONICS INDUSTRY

Thailand has achieved strong economic growth in recent decades. The Thai economy grew annually at an average rate of between 7.3 per cent and 7.8 per cent during the last 40 years. The gross national product (GNP) increased 35-fold between 1961 and 1998.

About 91 per cent of all the products of the electronics industry are exported. Accounting for 30 per cent of total exports, the electronics industry played a key role in this economic growth process.

Thailand's electronics industry has performed well in many respects:

- in 2000, the electronics industry contributed US$ 6.2 billion to Thailand's foreign exchange revenue;

- the number of persons employed in the electronics industry is predicted to rise from about 295,000 in 2001 to around 310,000 in 2005; and

- the electronics industry is one of the major recipients of FDI in the manufacturing sector.

The international competitiveness of the Thai electronics industry has largely been based on the acquisition and adaptation of foreign technology. This has mainly been achieved through appropriate incentives and policies.

12.3.2 POLICIES SUPPORTING THE ELECTRONICS INDUSTRY

Over the last four decades the Thai government has developed a variety of policies to develop the economy. We distinguish between overall development policies (Section 12.3.2.1), policies to attract FDI (Section 12.3.2.2), policies to attract technology (Section 12.3.2.3), and intellectual property rights (Section 12.3.2.4, see Figure 12.2).

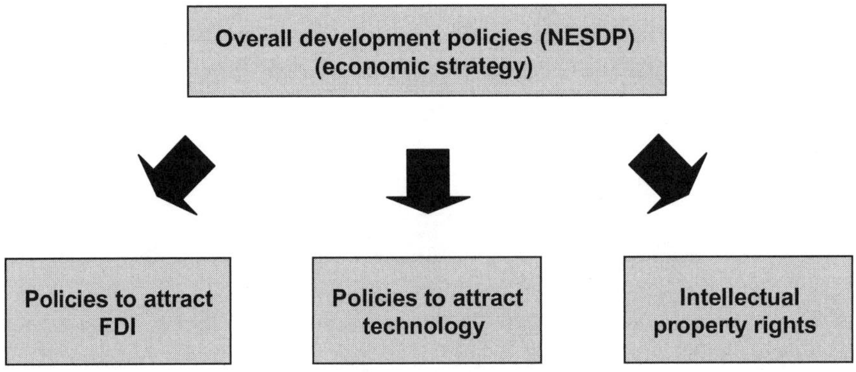

Figure 12.2   Thailand's development policies

12.3.2.1 Overall development policies (economic strategy)

Since 1961, the development policies of Thailand have been formulated in a five-year National Economic and Social Development Plan (NESDP). The NESDP is a strategy plan for the economic development of the country (see Chapter 8). Each plan contains detailed targets such as economic growth rate, inflation rate, export growth rate and the number of jobs to be created. It also explained how the targets should be achieved.

The electronics industry had four distinct development periods:

- 1961–1971 (first and second plan) – government policies aimed at promoting production of consumer products for the domestic market. Tax incentives and tariffs were structured to attract FDI in the manufacturing sector to supply the home market. Overall, this strategy only achieved modest growth owing to the limited size of the domestic market.

- 1972–1985 (third to fifth plan) – policies focused on export promotion, supported by improved access to the US market. The electronics industry was able to diversify its export product range and to increase production volume. Major world producers of electronic products established production facilities in Thailand.

- 1986–1992 (sixth plan) – Japanese, Korean, and Taiwanese firms relocated their production facilities to Thailand. These firms played a central role in the rapid growth of electronic exports. They also gave rise to the development of some supporting industries such as subcontractors and suppliers.

- 1993–present (seventh to ninth plan) – the policies sought to exploit the continuous expansion of the electronics industry.

The national development plans have had two major effects:

- first, the formulated a sound macroeconomic profile and enhanced the confidence and trust of foreign investors;

- secondly, they creating opportunities for export-oriented FDI and technology transfer.

12.3.2.2 Polices to Attract FDI

One instrument playing an important role in the development of the electronics industry is the Board of Investment (BOI). BOI is the Thai government agency responsible for investment policy advocacy, its implementation and investment promotion. BOI functions as a 'one-stop shop', providing investors with a wide range of incentives such as exemptions from and/or reduction of import duties, corporate income and other income-related taxes, credit assistance, guarantees, infrastructure support and services (see also Section 11.1).

The support offered to firms, such as low import tariffs on electronic equipment and parts needed to manufacture exports, was important in enabling the industry to compete internationally. The electronics industry accounted for between 18–40 per cent of granted incentives by BOI from 1998–2002. BOI has also introduced incentives for firms to engage in R&D, training and marketing. In addition, there are provisions to facilitate recruitment of foreign IT personnel to work in Thailand.

12.3.2.3 Policies to attract Technology

Several policies to attract technology were part of the national development plan.

- For instance, the fifth NESDP (1982–1986) included policies that specifically sought to promote transfer of foreign technology, develop science and technology manpower and increase R&D. As a result, several technical cooperation agreements were signed with other countries. For example, the Science and Technology Development Project, with technical assistance from the United States, had a budget of US$ 49 million to support technology activities over a period of seven years.

- Several institutes and agencies were founded to build research capabilities in the public sector and increase absorptive capacity (see Section 12.1). For example, the Ministry of Science, Technology and Energy was established in 1979 as the lead government agency for the planning and implementation of science and technology.

- In addition, the sixth plan (1987–1991) identified key technologies for the electronics industry such as computer-aided engineering and production management (see Chapter 8). It also sought to promote cooperation among the various government agencies involved in R&D and to improve the linkages between the R&D institutions and the industry. This facilitated the emergence of industry clusters and generating opportunities for spillovers (see Section 12.2).

12.3.2.4 Intellectual Property Rights

For the electronics industry, the need to protect designs of electronic systems is very important. The development of a new design requires substantial investment but it can easily be copied and built. It is for this reason that Thailand introduced legislation in 2000 for the protection of integrated circuit design for 10 years.

Similarly, software plays an important role. Thailand revised its Copyright Act in 1995 and in 2001 ratified the Berne Convention and its obligation in respect of copyright protection under the TRIPS Agreement. Thailand has therefore made great strides towards implementing an effective intellectual property rights system that seeks to meet the needs of the electronics industry.

12.3.3 TECHNOLOGY TRANSFER TO THE THAI ELECTRONICS INDUSTRY

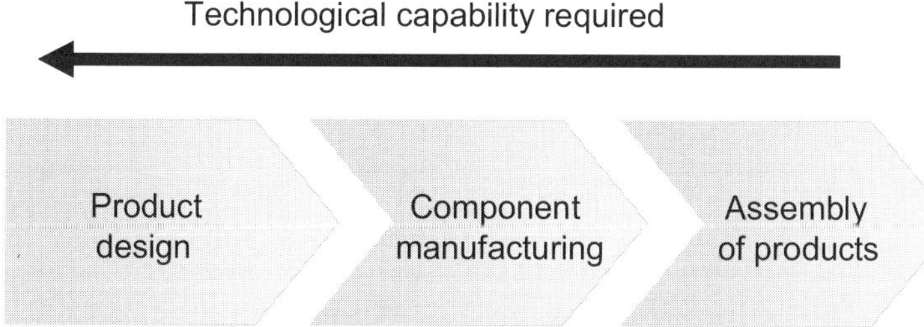

Figure 12.3   The production process of electronic goods

The production of electronic goods may be divided into three stages (see Figure 12.3):

- The *product design* requires extensive technical knowledge and investment in R&D to develop novel products and processes.

- The *manufacturing* of components requires capital-intensive investment for mass production.

- The *assembly* of the final products is capital and labour-intensive and requires lower skills than the other stages of production.

The Thai electronics industry started with the *assembly* of low-technology consumer products (radios and televisions) and has steadily developed into assembly and manufacture of high-technology products (microwave isolators, hard disc drives).

However, *manufacture and design capabilities* remain limited to a few sub-sectors. For instance, the hard disc drives sub-sector has acquired strong manufacturing capabilities but has not yet acquired research and product development capabilities. Similarly, the Thai semiconductor industry has some capacity in assembling but lacks manufacturing capacity and the ability to design new products.

This may be explained by the fact that technology spillovers in the electronics industry have largely occurred within the production networks of MNCs. As contract manufacturers, the domestic firms are more inclined to assemble or manufacture products required by the MNCs than develop novel products that compete with the established firms. This has limited the innovative activities to process upgrade and design.

Compared to other Asian countries, Thailand has a lower concentration of researchers and private sector expenditure on R&D per capita. The R&D performance reveals that government and higher education institutions account for more than 60 per cent of Thailand's national R&D expenditure.

In general, the performance of the industry in terms of physical productivity closely matches that of its Asian competitors. However, the value-added productivity in the electronics industry is lower than that of Singapore and Taiwan. This reflects the fact that the industry is largely at the assembly stage.

12.3.4 KEY LEARNINGS

Thailand's electronics industry has successfully acquired *assembly technology* and partly manufacturing technology. Key success factors for this development were:

- a clear overall strategy in the form of the NESDP;

- successful policies in attracting export-oriented FDI to the electronics industry, especially via the creation of the BOI as a 'one-stop shop'; and

- a variety of policies directly aiming to promote technology transfer, namely the building of industry clusters around MNCs.

However, the Thai electronics industry has not yet been very successful in acquiring design as well as R&D capacity. Some reasons are:

- Thailand still lacks absorptive capacity and has a lower concentration of researchers and private sector expenditure on R&D per capita than other countries.

- R&D activity is predominately taking place in the public sector while only a limited number of private firms undertake R&D.

- The second point is intensified by the fact that the electronics industry has only weak linkages to public research institutions.

Consequently, the government has developed several initiatives to encourage the electronics industry to move from assembly to the manufacture and design stages.

13

Technology Application

13.1 Introduction

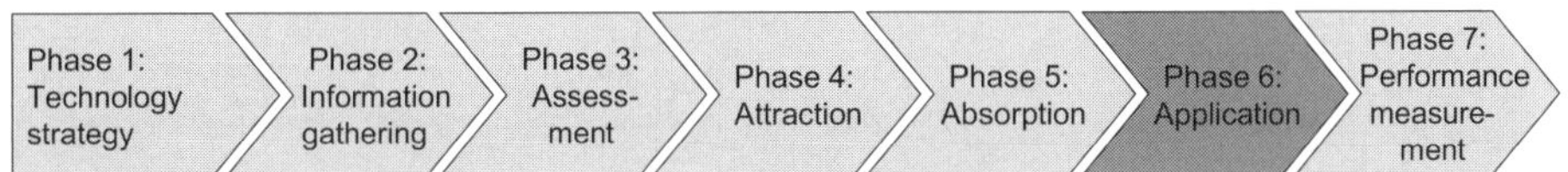

Figure 13.1   Technology application (phase 6)

The *application* of technology in a firm, industry and country formulates phase 6 of active technology management (see Figure 13.1). While attraction and absorption are necessary preconditions for this step, they do not yet guarantee the actual use of the acquired technology. In various cases promises of productivity gains are only partly realized, often because the social organization could not adapt to the use of new technologies (see also Cohen 2004: 147, 148).

The following factors generally affect the acceptance and applicability of new technologies (see Phillips 2001: 271, Cohen 2004: 150):

- The characteristics of the technology, for example observability of positive results, complexity, trialability, etc.

- The extent to which those who work in a company share the technology goals and values.

- The approach chosen for the introduction of a new technology.

- The general attitude and culture of the population of a country/region towards the introduction of new technologies and products.

- Closely related to the above, the way an organization, region and country is used to technological changes.

- Macroeconomic factors.

- Government regulations and legislation.

Some factors such as the culture and macroeconomic factors are outside the control of a private firm or public organization. Others, such as the approach chosen to introduce a new technology, can be influenced and need to be actively managed. Box 13.1 gives an example of an unsuccessful technology application.

BOX 13.1   FAILING APPLICATION OF BOILING WATER IN A PERUVIAN VILLAGE

Once, a social worker was sent to a small village in Peru to persuade women to boil drinking water as a disease preventive. The social worker was an unmarried woman from the city. The married village women were the authority figures who had to be persuaded to boil their families' drinking water. In the eyes of the married village women the social worker was not a credible advocate of the use of this new technology.

The social worker won the trust of two village women. One of these, however, was seen as a hypochondriac. The other was a migrant from a distant region who had never become accepted as a village insider. Neither woman was an effective opinion leader for the new technology. Further, the social worker was ignorant of local superstitions regarding the 'hot' and the 'cold' spirits inherent in all materials, including water. Her explanation of the reasons for the boiling of water conflicted with these traditions, confusing the village women. Finally, the impossibility of showing the effect of boiling water on the bacteria made it difficult to demonstrate unequivocally the health benefit of boiling. The application of the new technology failed and the social worker departed without having changed the behaviour or the health of the villagers.

Adapted from Phillips (2001: 269).

In order to provide you with knowledge of how the application of technologies can be facilitated, we have structured this chapter as follows:

- First, we address obstacles for the internal usage of process technologies in firms and possibilities to overcome these (Section 13.2).

- We then examine different options for the external commercialization of product technologies (Section 13.3).

13.2 Internal Usage: Process Technology

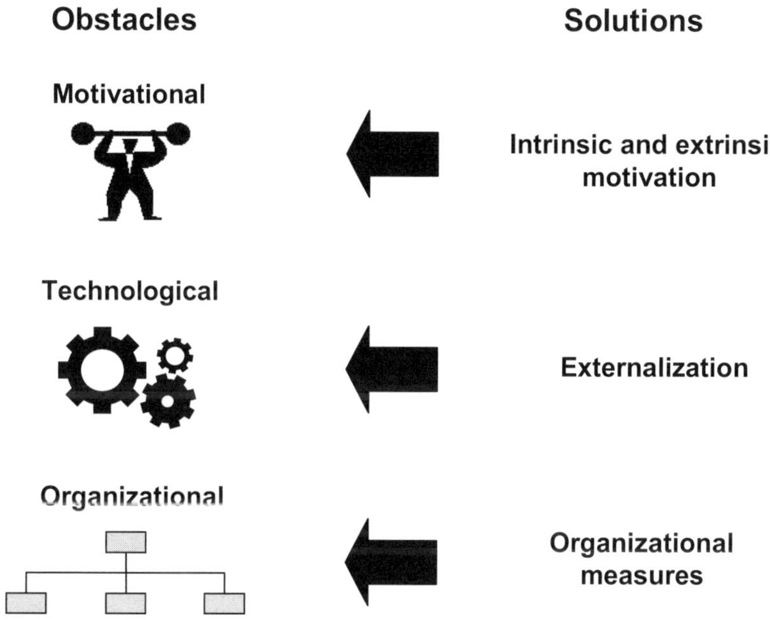

Figure 13.2   Obstacles and measures for technology usage

Process technology refers to the tools, materials, equipments and other technologies that directly affect how a firm develops, produces, or distributes the goods that it sells on the market (see Section 2.3.3). If a firm has absorbed a new process technology, it faces the challenge of promoting its application throughout the organization. Even the effective distribution of a technology in a firm via its internal communication network is only a necessary, not a sufficient condition for technology application. As shown in Figure 13.2, there are three obstacles for the adequate usage of absorbed process technologies (Al-Laham 2003: 355–357). Each obstacle requires different solutions.

The first problem is a *lack of motivation* by organizational members to actively use new technologies. Employees often feel uncomfortable having to ask for other people's knowledge and skills and see such asking as potentially harming for their own reputation. Fears exist that one's own knowledge gap might become visible to colleagues or superiors. Consequently, organizational members stick to their routines of daily business which form a type of protection mechanism against innovations. The more the employees of a firm stick to their routines, the more difficult it becomes to motivate them to apply new technologies.

In order to overcome this obstacle, managers must foster *extrinsic and intrinsic motivation* to use new technologies (see Osterloh and Frey 2002, Shane 2009: 341–344):

- *Extrinsic motivation* refers to the motivation of employees by tangible rewards (payment, promotions, punishment) or intangible rewards (praise or public commendation). For example, a superior could encourage the use of a new customer database by actually commending it. Or managers could grant payment bonuses to engineers which implement new production technologies. However, measures to extrinsically motivate employees should be used with caution. They can lead to perverted effects, such as employees using technologies solely to acquire compensation. Another consequence can be employees taking excessive risks in order to achieve the external reward or neglecting the importance of other activities.

- People act on *intrinsic motivation* when they engage in an activity for its own sake, without some obvious external incentive present. A hobby is a typical example. Intrinsic motivation may be influenced by shaping the culture of an organization. Norms and values reducing fears of the use of new technologies need to be strengthened by the leading management team. Leaders have to advocate experimentation and trial and error procedures as desirable activities. This involves some tolerance regarding mistakes and errors. In addition, the importance of new and innovative technologies for the success of the firm has to be continuously communicated. Also, the management needs to recruit personnel who are open and flexible to applying new technologies and show curiosity for new ideas.

A second problem for the application of new technologies is that of *technological obstacles*, for example the lack of user-friendliness of technologies. For some technologies, their positive effect might not be directly visible to their potential users: this was the case with the example in Box 13.1. The functions of the technology might be ill-documented, making it hard for employees to understand how certain procedures or techniques need to be performed. It may be that a technology in itself is overwhelmingly complex, requiring a relatively high absorptive capacity.

To tackle the problem of technological obstacles, *externalization* needs to be used to enhance the understandability and user-friendliness of new technologies. As shown in Chapter 4, externalization refers to the conversion of tacit to explicit technology and takes place when an individual is able to articulate the foundations of their tacit technology. For example, technical specialists need to document absorbed technologies in a straightforward manner and use everyday language. This will facilitate the application of new technologies for other employees. If necessary, the technology itself needs to be modified to enhance user-friendliness.

Finally, a lack of usage of absorbed technologies may be due to *organizational problems*. These can be caused by structural barriers such as a gap between internal technology suppliers and demanders. Lack of transparency on the availability of internal technology is another organizational feature which harms the application of technologies. In addition, strong organizational routines, inflexible rules and bureaucracy make the establishment of new technologies difficult.

In order to overcome these obstacles, *organizational measures* have to be applied. Overly excessive bureaucratic rules and regulations need to be downsized to give room for the 'trial and error' of absorbed technologies. Additionally, mixed teams containing technology suppliers and users are required. Because many process technologies affect the entire value chain, managers should also make use of cross-functional teams, involving design, production, sales and marketing specialists (Shane 2009: 347). Organizational structures may need to be altered, pushing towards a project-based organization instead of permanent tasks. In addition, a matrix structure with employees reporting to a functional and a project team superior can also enhance the implementation of new process technologies (Shane 2009: 349).

13.3 External Usage: Product Technology

Product technology refers to the knowledge related to a certain product a firm is selling for the benefit of customers or clients. If a firm has absorbed a new product technology, it needs to apply this technology externally, commercialize and capture value from it. Governmental organizations or consultants may assist firms in this process.

13.3.1 OPTIONS FOR COMMERCIALIZATION

The following options for the external usage of absorbed product technologies exist:

- A firm may decide *not to commercialize* an absorbed technology. Such a decision may be based on a limited market potential for the technology at the current state. Also, some technologies may only be viable for internal application.

- A firm may sell *the absorbed technology* to others, for example in the form of licenses. Licensing may occur because firms lack capital for marketing a technology, want to avoid risks, or are not in possession of a distribution network. For example, the Michigan State University licensed its patent to 'shikimic acid', a basic ingredient in the flu drug Tamiflu, to Roche Pharmaceuticals; Intel licensed its instruction set for producing microprocessors in personal computers to AMD (Shane 2009: 317, 318). Licensing a technology to another firm includes the commercialization of documents and know-how and can involve training the other firm's experts (Chen 2006: 298). Licensing can offer considerable financial returns, especially if a company wants to avoid the costly and difficult process of creating manufacturing and marketing capabilities. For example, IBM estimates it would have to make an additional US$20 billion in sales per year to receive the same net income as the US$ billion it earns from royalties on licensing its technology (Shane 2009: 318). However, licensing has the disadvantage that a firm has less control over the licensee than if it had set up its own production and sales facilities. Furthermore, if the licensee is very successful the firm has given away some of its profits, and when the contract ends, the company might find that it has created a competitor (Kotler 2000: 377). Selling to another national or foreign firm of a developing country is considered a special form of South–South technology transfer (see Box 13.2).

BOX 13.2   SOUTH–SOUTH TRANSFER OF TECHNOLOGY

Developing countries increasingly transfer technology among themselves. In the past, a number of South–South initiatives have occurred through regional and bilateral agreements.

For example, China and Brazil agreed in 1989 to develop two remote sensing satellites through the China–Brazil Earth Resources Satellite (CBERS) Programme. The programme pools the human and financial resources of both countries to establish a remote sensing system that is internationally competitive. The National Institute for Space Research (NIPE) in Brazil and the Chinese Academy of Space Technology (CAST), are the lead agents in this programme. CBERS launched its first remote satellite in 1999 aboard a Chinese rocket from the launch centre in Taiyuan after 13 years of cooperation. The second satellite was launched in 2001. China bore 70 per cent of the cost while Brazil covered 30 per cent. Brazil is responsible for the development of the high-resolution cameras while China is responsible for the application platform.

Another example for South–South transfer is the International Centre for the Advancement of Manufacturing Technology (ICAMT) based in Bangalore and serving as a gatekeeper of appropriate and emerging manufacturing technologies. ICAMT was established by UNIDO in cooperation with the government of India to promote manufacturing technologies for industrial competitiveness in developing countries and to foster international cooperation. The government of India, as the host country, provided the premises and US$1.3 million while UNIDO also allocated some funds for this purpose. The centre promotes both North–South and South–South technology transfer. It is involved in the generation and transfer of Indian and Asian technologies to other developing regions, mainly Africa, Latin America and the Caribbean. The centre has also initiated new projects in Kenya, Malawi, Mozambique, Uganda, Zambia and

> Zimbabwe to provide manufacturing technologies and expertise on low-cost housing and encourage technical cooperation among developing countries in Asia and Africa. ICAMT has also taken part in exhibitions in Africa, through the Asia–African Technology Partnership Forum.
>
> Adapted from UNCTAD (2004: 17–18).

Another possibility for commercializing a foreign technology consists in the foundation of a *joint venture* with a partner firm. As mentioned in Chapter 5, a joint venture is a strategic alliance between two or more parties to undertake economic activity together. For example, Samsung formed a joint venture with Sony to make flat-panel televisions (Shane 2009: 316). A firm in a developing country may search for another party, create a new entity, contribute equity and share revenues, expenses, and control of the enterprise. Such a venture can be suitable if a firm does not have the necessary capital, marketing experience or lacks a distribution network for the new technology (Shane 2009: 316). However, a joint venture bears the risk that partners might disagree over investment, marketing and other policies (Kotler 2000: 377).

Finally, the firm may make *direct investments* and introduce the new technology alone in the market. This way a firm can fully exploit the rents deriving from its technology. However, firms should know their target customer, have experience in the market and be able to find distribution channels to bring their products to their customer (Chen 2006: 303).

13.3.2 DEVELOPING A MARKETING STRATEGY AND A MARKETING MIX

If a firm has decided to externally commercialize a product technology via a joint venture or own investments, it needs to market the product. To do so, two elements are required:

- The *marketing strategy* consists of market segmentation as well as the selection of a target market. It focuses on the long-term positioning of a product technology in the market.

- The *tactical marketing* refers to the actual marketing mix consisting of activities such as price, promotion, distribution channels (place) and product characteristics.

Figure 13.3   The marketing process

13.3.2.1 Marketing Strategy

The marketing strategy for a technology is derived from the overall strategy of a firm (see Figure 13.3). While the overall strategy of a firm consists of a firm's mission statement and its strategic themes, a marking strategy consists of a market segmentation and the selection of a target market for a technology (see Kotler 2000: 256–278, Baines, Fill and Page 2008: 217–251):

- A *market segment* consists of a large identifiable group within a market with similar wants, purchasing power, geographical location, buying attitudes, or buying habits. Segmentation criteria such as regions, age, or lifestyle should fulfil the standards of homogeneity (people within the segment are similar) and heterogeneity (customers between segments are different). Segmentation is an approach to marketing a technology or a technology-based product in-between mass marketing and individual marketing. Each segment's buyers are assumed to be quite similar in wants and needs, yet no two buyers are exactly alike. A market segmentation offers several benefits such as the possibility to create a fine-tuned product or service, a price appropriate for the target audience, or an adequate choice of distribution and communication channels.

- Ones a firm has defined its market segments, it needs to decide how many and which ones to *target*. Each market segment needs to be evaluated: firstly the firm must ask whether a potential segment is attractive in terms of size, growth, profitability, or risk. Secondly, the firm has to consider if investing in a certain segment is congruent with the company's resources and objectives. A firm should then target the segments with the highest attractiveness and congruence to firm's goals and resources. Box 13.3 gives a fictive example of market segmentation and targeting in a developing country.

BOX 13.3   MARKET SEGMENTATION IN ECUADOR

In a fictive case, an Ecuadorian company has absorbed technology for a solar energy-based boiler. The firm wants to sell the boiler to households in Ecuador. To develop a systematic marketing plan, a market segmentation is developed. The following segmentation criteria are used:

- Geographic: where do the potential customers live (large cities, towns or villages)?
- Demographic: what income does the household have (large, medium, small)?

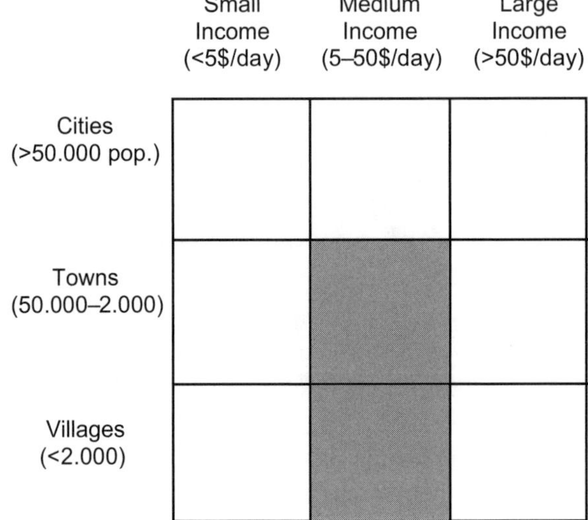

The two criteria were selected because they resulted in a useful segmentation:

- First, consumer preferences were expected to be equal within segments (homogenous). For example, all households in villages with medium income were expected to have similar needs and buyer customs.
- Secondly, customer's needs were different between distant incomes and regions (heterogeneous).

The firm views the medium income segments as attractive because purchasing power is sufficient to buy the boiler and households in this segment are perceived to have an actual need for boilers. Because the boilers need to be used outside, cities are dismissed and towns and villages are targeted.

13.3.2.2 Marketing Mix

The tactical marketing of a technology or its product consists of the formulation and implementation of the marketing mix. A marketing mix, also called the four Ps, consists of the following components (see Kotler 2000: 87, 379–386, Baines, Fill and Page 2008: 355–423):

- For the *product*, a firms needs to define the product quality, design, features, branding and packaging for each segment. Generally, a firm has to consider whether there should be modifications for the various market segments or if the product technology should be introduced without any changes.

- Secondly, firms need to define the *price* for a product technology. Prices may be based on costs, demand, competitor pricing, or value. For example, a firm may optimize its revenues by estimating impact of a change in price on demand. Additionally, a company has to consider whether it should set a uniform price or adapt the price in each market segment.

- The *place* includes the various activities the company undertakes to make the product accessible and available to target customers. Firms must understand the various types of retailers, wholesalers, and physical distribution firms relevant for their technology.

- Finally, the *promotion* of a product technology includes all the activities the company undertakes to communicate and promote its products to the target market.

Performance Measurement

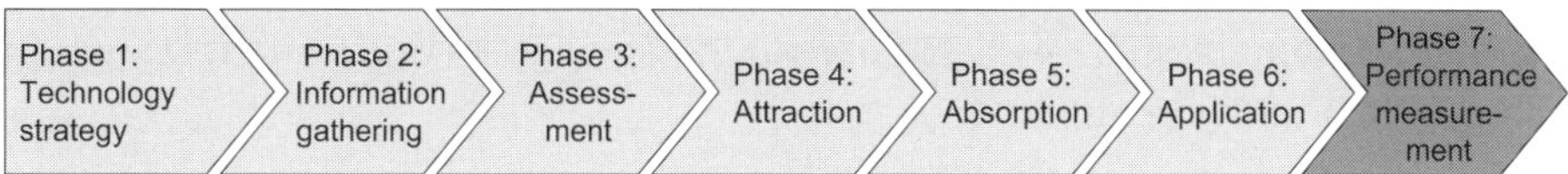

Figure 14.1   Performance measurement (phase 7)

Performance measurement is the last phase in the technology management process (see Figure 14.1). Generally speaking, *performance measurement* refers to the checking of an organization's or activity's progress towards planned goals. Performance measurement is therefore essential to determine the success of the whole technology strategy, to make necessary adjustments and to learn from past experiences. It involves the systematic measurement of whether the strategic goals were achieved and, if they were, how efficient and economical the process was. This information can then be incorporated into the design of a new technology strategy and the implementation of all the other activities of the technology management process.

This chapter is structured as follows:

- First, we explain the benefits of performance measurement (Section 14.1).

- We then introduce the different types of performance measurement in technology transfer (Section 14.2).

- We concluded by providing you with indicators to monitor the technology transfer facilitated by your organization (Section 14.3).

14.1 Benefits of Performance Measurement

An increasing number of firms and public organizations apply performance measurement to improve their functioning. In the private sector, performance measurement is used to improve competitive advantage by monitoring costs, effectiveness, customer satisfaction and return on investment. In the public sector, managers use performance measurement to analyse the efficient use of public resources.

In the context of technology transfer, performance measurement enables the management of an investment promotion agency or similar organization to achieve the following aims:

- Measure progress against the *goals of the technology strategy*. Tracking and measuring the performance of technology transfer enables managers to assess the progress against the goals formulated in the technology strategy. In this way managers acquire the knowledge for necessary strategy modifications.

- *Learn from experience* of past projects. The data of the monitoring process allows managers to learn from experience. This requires an exact documentation of project-related costs and results.

- Establish a *feedback culture*. The use of performance measurement systems shapes the way managers in organizations act, think and communicate about goals, processes and outcomes. A monitoring system facilitates the development of a common language about central issues of technology transfer.

- Collect information for *promotional campaigns*. Performance measurement of technology transfer generates useful information for promotional campaigns. If an investment promotion agency can report on successful technology transfer projects, it excites the interest of stakeholders in further projects.

- Collect *generally useful data* on technology transfer. Performance measurement produces data in a format that is useful to other stakeholders, for example, other governmental organizations, the media, potential investors and local companies.

- Prove the efficiency of an organization to its *clients*. Performance measurement shows an organization is publicly accountable. It enables managers to demonstrate their technology transfer achievements in a concrete form and to prove the efficient use of funds to public or private clients.

14.2 Types of Performance Measurement

Although widely used in theory and practice, the term 'performance' enjoys a diverse usage. Performance simultaneously refers to the action, the result of the action and the success of the result compared to some benchmark. One may define performance as doing today what will lead to measured outcomes tomorrow (Lebas and Euske 2002: 68).

The creation of performance by an organization may be visually illustrated as a tree (Figure 14.2). Three levels are interrelated to each other (see Lebas and Euske 2002: 79):

- *Results* are the fruits of the tree. These fruits are the technological outcomes and offer value to the stakeholders of an organization. They include figures such as the number of patents, the financial benefit (cash flow) of commercialized technologies, or the amount of R&D-related FDI. A variety of measures has to be included to get a complex and comprehensive picture.

- These outcomes are the results of the *business processes* constituting the trunk of the performance tree. Business processes have to be monitored to deliver what the stakeholders want. For example, in phase 4 (attraction), a measure could be the amount of tax reductions granted to foreign firms transferring technology to the host country.

- The quality of the results depends on the *premises* of the processes. Premises consist of the assumptions and goals of the technology strategy and are illustrated as the nutrients of the soil in Figure 14.2. Premises need to be monitored in order to judge whether they are still valid or not. Assumptions include estimates on the prior technology basis or the human resources of a region (see Chapter 8).

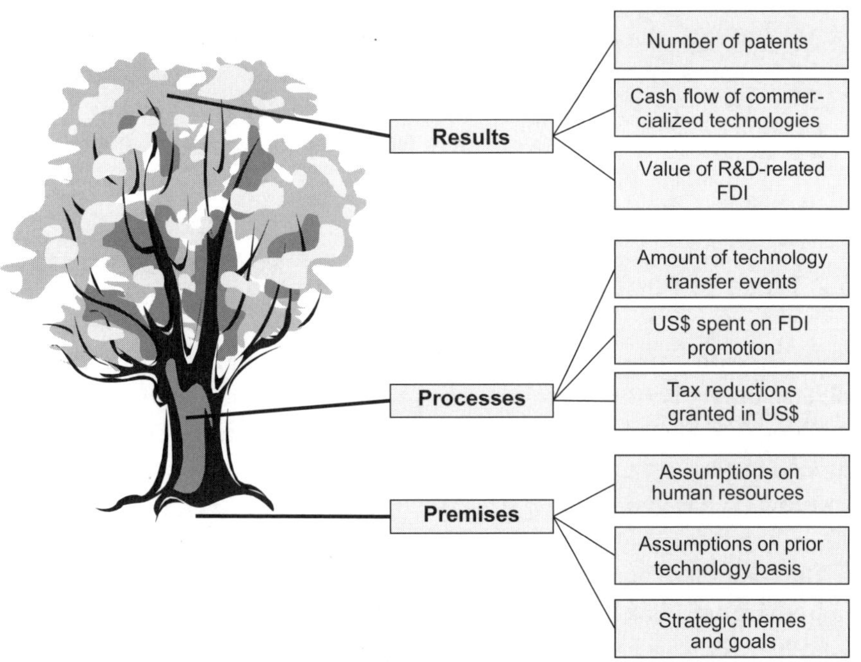

Figure 14.2   The tree of performance measurement

Describing performance measurement as a tree illustrates that outcomes and actions do not occur in the same time frame. Performance measurement needs to be conducted for long time periods. For each level of performance measurement, it is important to remember that qualitative and quantitative indicators are mere surrogates of performance. They should not be mistaken for performance itself (Lebas and Euske 2002: 74).

In accordance with the three levels of performance, we distinguish between three types of performance measurement (Figure 14.3, see also Müller-Stewens and Lechner 2005: 694–697):

- The *control of premises* is necessary to create awareness on changes in the opportunities and threats of the technology strategy (Chapter 8). This form of measurement asks whether the assumptions of a technology strategy are still valid. It has to be measured if the internal and external factors of a technology strategy have changed. For example, are the estimates of the prior technology basis still correct or do we need to make adjustments? Have the technological

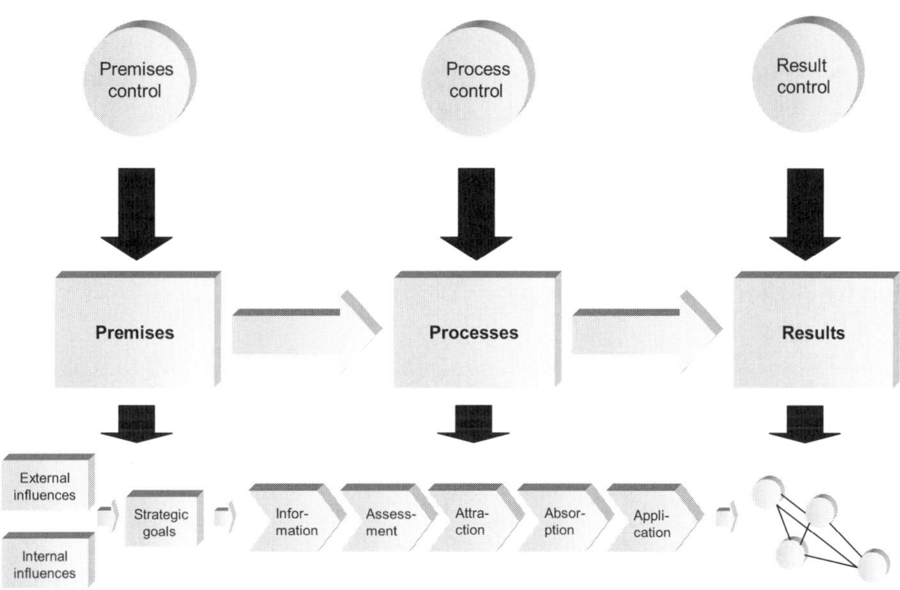

Figure 14.3   Three types of performance measurement

trends changed? Are there modifications in our human resources situation? Has the demand for a certain technology changed? What consequences do these changes have for our vision and our strategic goals? The control of premises is necessary because the results of a technology strategy might not be observed until some years later. However, the management of an organization needs continuous feedback to make the necessary strategy adjustments in time.

- The *control of processes* deals with the actions undertaken to achieve the strategic goals. In this context managers should monitor whether the indicators developed have been implemented correctly and in time. All phases of the technology management cycle need to be monitored. For example, it might be asked whether the gathering of information (Chapter 9) was actually performed or incentives for the attraction of technology-related FDI have been passed (Chapter 11). It should also evaluate whether the technology strategy was accepted by the internal and external stakeholders of a firm.

- The *control of outcomes* focuses on the achievements of a technology strategy. Even if premises have proven to be correct and processes were implemented the right way, the results of a technology strategy might not be consistent with the expectations. Typical questions are: have we come closer to the achievement of our vision and our strategic goals? Have we increased the technology basis of our country/region? Were the technology transfer projects successful?

We will now illustrate how you can develop concrete indicators for each type of performance measurement.

14.3 Measuring Technology Transfer

The creation of performance indicators in general faces three obstacles (see Austin and Larkey 2002: 328, 329):

- The problem of *measurability* refers to the difficulty of quantifying relevant aspects of the work observed. In situations where direct measurement of results seems impossible, less direct proxy indicators are often sought. For example, software quality is difficult to measure. Defect counts are often used as a proxy indicator but in reality are only suggestions of quality as some customers define it.

- The problem of *attributability* is the degree to which a measurement can be attributed to an individual, a group, or a process. Measuring something without knowing its relationship to an individual, group or process is not very useful. For example, the achievement of R&D is difficult to attribute to individuals as this is usually the result of a team effort.

- *The problem of evaluability* is the extent to which a measurement can be normatively judged. For example, US$50 million of R&D-related FDI inflows per year do not yet give any information as to whether this amount is sufficient or insufficient. Standards and benchmarks usually assist with evaluability issues.

We will now explain how performance indicators may be created for premises control, processes control and result control.

14.3.1 MEASURING PREMISES

Technology transfer is a long-term business. Few strategic themes lead immediately to new technologies or to measurable results. It can take years to identify, attract and absorb foreign technologies. The long time that it takes for results to materialize means that agencies cannot wait for results to evaluate their strategy. Instead, the analysis of premises and strategic goals has to be performed continuously. For instance, the demand for a technology might suddenly change, making part of the technology strategy obsolete. For this reason one should assign an observable indicator that can be continually monitored to each opportunity, threat, strength and weakness of the technology strategy. For example, the human resources of a country may be measured by the percentage of university students or the percentage of the population that can read and write. In addition to quantitative, qualitative indicators need to be applied to monitor changes in premises. Strategy modifications should be initiated if critical values of premise indicators are exceeded. The best way to do this is to conduct a strategy workshop monitoring changes and modifying the technology strategy on a regular basis.

14.3.2 MEASURING PROCESSES

In addition to premises, actual processes of technology management have to be monitored. The performance indicators used for process monitoring should be developed in accordance with the actual activities performed. As processes vary in every organization, general indicators cannot be prescribed. Instead, managers need to adjust indicators to their specific business processes. Table 14.1 gives an overview of potential activities and possible performance indicators. In the phase of information gathering on technologies, for instance, activities such as visiting trade fairs could be measured by the number of fairs visited or the number of contacts established during such events.

Table 14.1   Potential indicators for business processes

Phase of technology management	Activity	Potential performance indicators
Information gathering (phase 2)	Visiting trade fairs	Number of trade fairs visited Contacts established at trade fairs
	Establish contacts to transfer intermediaries	Number of contacts in database Quality of contacts
	Formulate profiles on interesting technologies	Number of technology profiles Content of profiles
	Establish contracts with universities and public research institutes	Number of contacts in database Quality of contacts
Technology assessment (phase 3)	Conduct Delphi survey on technologies	Delphi surveys conducted
	Develop scenarios	Scenarios developed
	Calculate discounted cash flow	Discounted cash flow calculated
Technology attraction (phase 4)	Promote fiscal incentives for FDI	Promotional events performed Direct mailings Webpage hits Information brochures distributed
Technology absorption (phase 5)	Facilitate technology spillover via vertical linkages with local suppliers	Match-making events organized Contacts established to MNCs supply-chain managers
Technology application (phase 6)	Push marketing of absorbed technologies of local firms	Marketing seminars performed with local firms

14.3.3 MEASURING RESULTS

While process indicators can be directly derived from the activities of investment promotion agencies, the measurement of results is more difficult. It is obvious that an investment promotion agency's effectiveness on technology transfer is ultimately measured by the amount of new technology brought into a country or region. However, as technology is tacit as well as explicit and is often exclusively stored in the minds of organizational members, there are problems of measurability and accountability. Reliable indicators to directly assess the value of an acquired technology do not yet exist.

As with process indicators, result indicators have to be related to the desired outcomes. These are usually formulated with the technology strategy. Therefore, result indicators depend on the specific strategy of an investment promotion agency. For example, if one goal of technology strategy is to increase patents in the agricultural industry, then published data on patents in this field are indicators on the effectiveness of technology transfer.

In order to facilitate the process of finding indicators we want to mention some that might be of general concern.

Amount of *R&D-related FDI inflows*: as mentioned in Section 2.5, innovative skills constitute the highest level of technological capabilities. Consisting of formal R&D, innovative skills are necessary to generate new technologies and to keep pace with technological development. For FDI inflows, data is regularly collected by international organizations such as UNCTAD. R&D-related FDI flows are often declared separately and may be used as a performance measure for technology transfer. However, it should be kept in mind that formal R&D represents only one part of the technological activity in a country. As R&D transfer is the most advanced form of technology transfer, this indicator might not be very suitable for less developed countries which primarily transfer lower-level capabilities (for example assembly or manufacturing capabilities).

Number of *R&D-related jobs created* from FDI: to measure the amount of technology transferred on the level of innovative skills indirectly, one may also use the number of jobs created within formal R&D. While this indicator is a good contemplative variable for R&D-related FDI inflows, it faces the same problems.

Number of patents granted to foreign firms: a modern patent provides the right to exclude others from using or selling an innovation and is obtained by filing an application with a patent office. Extensive patent data exists for the patents filled at the European Patent Office or the United States Patent and Trademark Office. The number of patent applications from developing countries to offices gives an indicator of innovative capabilities of developing countries. The share of patents granted to foreigners in these countries (for example to MNC subsidiaries) can then be used as an indicator of the technology transferred by foreigners to developing countries (see UNCTAD 2005a: 134). However, patent indicators face the same problems as the R&D measures because they only focus on the highest level of technological capabilities.

Amount of training hours of personal in FDI-related enterprises: as we have shown previously, technology is often transferred through personal interaction. Companies regularly use training sessions as mechanisms for technology transfer. The number of training hours constitutes a good proxy variable for the amount of technology transferred to a region. Indicators might be the training hours offered by a MNC to its employees in a subsidiary, to employees in a joint venture or to employees of a local supplier. However, two problems exist:

1.   Training hours only capture one mechanisms of technology transfer; and

2.   the hours trained do not yet give any indication of the type and the quality of the technology transferred.

Productivity growth in FDI-related firms: in Chapter 3 we saw that the productivity of an economy or firm is increased by technology. Productivity growth may therefore be used as a proxy of technology transfer. For instance, labour productivity is typically measured as output per worker or output per labour-hour. Measuring productivity gains has the advantage that all levels of technological capabilities are addressed. Though national data on productivity gains is widely available from the national accounting system, plant-level information might be hard to obtain. Also, productivity gains in local supplier firms, joint ventures or wholly owned subsidiaries might be due to factors other than technology transfer. For example, national consultants, local innovations, or national deregulation might also increase productivity. This increase is not equivalent with international technology transfer from FDI.

Increase of wages in FDI-related firms: because productivity gains are difficult to measure directly, wages might be used as a proxy variable. The idea is that the wages of skilled and unskilled workers increase in relation to the productivity growth of a firm (Saggi 2000: 22). Foreign affiliates are often able to pay higher wages than local firms because they have a higher productivity due to an advanced technology. While wage growth is easy to capture, measurability problems exist. Wage increase might not only be due to newly applied foreign technologies but also to other factors such as labour union activity or labour shortage.

Number of *university graduates employed* in FDI-related firms: new technologies often demand highly qualified employees. This implies that the amount of technology transferred might be measured by the number of university graduates employed in FDI-related projects. Nevertheless, the increase of skilled labour demand cannot exclusively be attributed to technology transfer. Factors such as a sound economic development or national innovations could be other determents.

Discounted cash flow: commercialized technologies create financial value by generating cash flows. The discounted cash flow generated by an attracted technology can therefore be used to estimate outcomes of technology transfer projects. In Section 10.2.5 we presented the formula for the calculation of discounted cash flows. While this financial indicator is a good proxy for the market value of a technology, it is not applicable for technologies that have not been externally commercialized. In addition, the effect of a new technology on the cash flows generated from a product might not always be clear.

Table 14.2 gives you an overview on the potential indicators for results of technology transfer.

It is difficult to measure the exact impact of an investment promotion agency or silimar organization. Much investment occurs without the involvement or even prior knowledge of agencies. Using nationwide data for the indicators described in Table 14.2 would therefore overestimate the impact of an investment promotion agency. On the other hand, collecting data from conducted projects does not take into account indirect effects of an investment promotion agency. For instance, some firms might have invested due to the promotional activity of an agency without being involved in a project. Measurement is also made difficult if other organizations played a role in attracting a particular business.

No method of data collection is capable of solving this problem. Therefore, the most exact measure is to focus on the direct impact of an agencies activity and measure project involvement only. National figures may be used as contemplative data.

Part or all of the information sought may be provided by the investor when seeking investment approval, when obtaining a license or another permit, or in the course of negotiating for government incentives. If additional information is needed, the agency may contact the investor directly.

It is also desirable to catch spillover effects by capturing figures of local suppliers receiving technology. For example, investors may provide figures on training hours conducted with local suppliers.

All of this information should be collected on a regular basis, preferably at the time that the investment occurs. It will be difficult to go back and to collect this data at a later date.

Table 14.2   Potential indicators for results of technology transfer

Type of performance indicator	Potential sources	Advantages	Disadvantages
R&D-related FDI inflows	• Information from projects with clients • National accounting system • Central bank • International organizations	• Easy to acquire • Relatively precise measure of R&D • Comparable to international data	• Technological output for a US$ of R&D varies • Large time lags • Measures only transfer on highest capability level
R&D-related jobs created from FDI	• Information from projects with clients • National accounting system • Central bank • International organizations	• Easy to acquire • Relatively precise measure of R&D • Comparable to international data	• Technological output for a person employed in R&D varies • Large time lags • Measures only transfer on highest capability level
Number of patents granted to foreign firms	• Information from projects with clients • International patent offices • International organization	• Relatively easy to acquire • Relatively precise measure of output of R&D	• Includes only patentable technologies • Patents not always commercially beneficial • Measures only transfer on highest capability level
Training hours of personnel in FDI-related firms	• Information from projects with clients	• Addresses all levels of technological capabilities • Estimate of increase in personalized technology	• Relatively difficult to acquire • Captures only one mechanism of technology transfer • No indication on quality of technology
Productivity growth in FDI-related firms	• Information from projects with clients • National accounting system. • International organization	• Addresses all levels of technological capabilities • Relatively good proxy for general technology growth	• Difficult to acquire • Productivity gains might also be due to other factors

Table 14.2   Potential indicators for results of technology transfer *concluded*

Type of performance indicator	Potential sources	Advantages	Disadvantages
Increase of labour wages in FDI-related firms	• Information from projects with clients • National accounting system. • International organization • Employer associations	• Addresses all levels of technological capabilities	• Wage increase might be due to other factors than technology transfer
University graduates employed in FDI-related firms	• Information from projects with clients	• Proxy for state of technological development	• Number of university graduates employed also depends on other factors
Discounted cash flow	• Information from projects with clients	• Represents market value of technology	• Only applicable for externally commercialized technologies • Effect of new technology on cash flow not always clear

15

Integrated Case Studies

15.1 Introduction

As we have shown, the transfer of knowledge and technology from one part of the world to the other is one of the most important driving mechanisms for global economic development. Today we see many highly developed and prosperous countries that were quite poor only 50, 40, 30 or even 20 years ago. Examples include Korea, Malaysia, Taiwan, or Singapore in South East Asia, Costa Rica or Chile in Latin America and Ireland or Estonia in Europe.

In this chapter we discuss cases in which FDI and the associated technology transfer have been of key importance in transforming economies and generating a competitive advantage for a nation. We have selected these examples because they stretch over distant time periods and different regions. They also focus on very different industry sectors, covering agriculture as well as manufacturing and service-related technologies. There are some key success factors that play an important role in every single one of these case studies:

- The government understood that to speed up the socio-economic development of the country it was of the utmost importance that it open itself up to FDI. In addition, the respective governments have shown great commitment and become very active in implementing measures for creating a business environment attractive to foreign investors. Crucial measures included reducing administrative barriers, accelerating processes for setting up new businesses, enhancing the protection of property rights of investors and providing adequate incentives (see Chapter 11).

- The different governments made the right strategic decisions in attracting investments with types of technologies that were appropriate and compatible to the available resources, the technological capability and the economic and social characteristics of the country (see Chapter 8).

- The respective governments have been wise in taking measures for the effective absorption and application of such technologies, for example by facilitating the emergence of industry clusters or supporting the development of qualified human resources (see Chapter 12 and Chapter 13).

Beyond these general issues, the importance of key success factors changes from case to case and depends on the specific characteristics of each country. While in one country the existence of a well-developed educational system is of dominant importance, in another the formation of industry clusters and the associated spillovers led to a successful absorption of new technologies.

In order to illustrate these different factors we describe the following case studies:

- Our first case study is historical and refers to the attraction of agriculture and agro-processing know-how by Russia in the seventeenth century and beyond (Section 15.2).

- The second case study is equally historical and describes the transfer of cotton and textiles technology to Egypt at the beginning of the nineteenth century (Section 15.3).

- Our subsequent case takes place in Chile's wine industry. It illustrates the importance of leveraging on a country's internal strengths in enhancing its production and marketing capabilities.

- The fourth case focuses on the Malaysian automotive industry and its role in the economic and technological progress of the country (Section 15.5).

- We then turn to Eastern Europe to illustrate the importance of internal technology transfer from MNCs to their local subsidiaries in the Estonian banking industry (Section 15.6).

- Finally, the case of the Irish software industry stresses the significance of investment in human resources and the educational system to attract and absorb technologies in the information and communication technology (ICT) sector (Section 15.7).

It is noteworthy that each case needs to be seen in the context of its respective time period. Countries might be successful in acquiring FDI and technology at one time but face difficulties in another due to changes in the environment, their internal factors or shifts in policies as the case of Ireland shows.

15.2 Russia: Agriculture and Agro-processing Technology

15.2.1 INVESTMENT PROMOTION UNDER CATHERINE II (1729–1796)

The empress Catherine II (1729–1796), also known as Catherine the Great, ruled Russia from July 1762 until her death in November 1796. Under her rule, the Russian Empire expanded, modernized its administration and started to develop along Western European lines. Catherine revitalized Russia, which became recognized as one of the great powers of Europe (Alexander 1988). Among her main successes was establishing a sound foreign policy and useful international cooperation. During her reign Catherine extended the borders of the Russian Empire southward and westward to absorb among others the Crimea, right-bank Ukraine, Belarus and Lithuania at the expense of two main powers: the Ottoman Empire and the Polish–Lithuanian Commonwealth. During her reign she added about 500,000 km² to Russian territory (Alexander 1988).

One of her remarkable achievements was the attraction of foreign investment and technology, particularly in *agriculture and food processing*, at that time a core sector of the economy. In this respect, Catherine became one of the first systematic foreign direct investment promoters in known history. After assuming the throne in 1762 she issued a manifesto, published in all European cities, which invited foreign farmers to come and establish new farms in the steppes newly conquered by Russia. Incentives provided were (Catherine II 1763):

- coverage of travel expenses;

- thirty years tax exemption;

- interest-free loans;

- freedom to change the business, provided foreign farmers could find somebody else for the farm; and

- equal treatment to Russian farmers.

The parallel to modern policy incentives to attract FDI and technology is striking (see Chapter 11). Catherine's objective was to attract people to the thinly populated regions of the country and bring the necessary technological expertise in order to increase agricultural production. During her reign, about 25,000 people, mostly German-speaking, moved to the Volga region. Among them were several hundred Swiss (see Bühler et al. 1985). It cannot be easily differentiated to what extent these people can be referred to as immigrants or as investors. Some had brought capital, technological expertise and know-how as well with them, as their instruments. Some were simple poor labourers, but many of them became wealthy farmers who reinvested their revenues into expanding their business. About 75,000 people moved to South Russia. Among those were Romanians, Bulgarians, Germans, Italians, French, Swedish and Swiss. The immigrants in the agricultural sector can be divided in two categories:

- the *settlers in land cultivation colonies* that started coming to Russia in the second half of the eighteenth century and in the first decades of the nineteenth century; and

- the *cheese makers* who followed later and reached the highest immigration flow in the years just before the First World War.

The dairy and meat product processing expertise and technology brought by the newcomers from alpine central Europe were particularly valuable because the increasing concentration of population in large agglomerations at that time needed new product designs and more efficient production and distribution systems.

15.2.2 IMMIGRATION AND INVESTMENTS AFTER CATHERINE II (1796–PRESENT)

Czar Alexander I, grandson of Catherine II, came to power in 1801 and further encouraged colonization. In his era, a group of 240 immigrants from the Canton of Zurich in Switzerland moved to Grim and established, together with German immigrants, the 'Zurich Valley'. Their main occupation was wheat cultivation and wine production. In 1817 there was another wave of immigration to Russia, this time to the Caucasus. Swiss and south German farmers left their countries, manly because of the famine caused by the potato crop failure and the influence of religious sects (Bühler et al. 1985).

Between 1795 and 1914 about 300 Swiss cheese-makers went to Russia. Most of them came from the Bern Highlands. The reason for this was a major recession in alpine cheese-making. The Swiss immigrants produced Emmentaler cheese in Russia (Bühler et al. 1985) (it is still called 'Swiss cheese' in Russia today). Well-established dairy producers brought workers from Switzerland. Sons took over the business of their fathers and thus dynasties of Swiss cheese-makers were established (Karlen, Wittwer, Stucki, Ammeter, etc.). Centres of milk economy were the regions of Smolensk, St Petersburg, Finland, Baltikum and the Trans-Caucasus. In some of these regions about half of the cheese factories were in Swiss hands. However, Russian producers entered soon into the diary business. Horizontal technology spillovers from Swiss to Russian cheese producers occurred (see Section 6.2): in 1894 only 15 per cent of the cheese factories remained in the hands of Swiss (Bühler et al. 1985). Nevertheless, Swiss cheese-makers were forerunners of the Industrial Revolution because they made an important contribution in transferring agro-technology and integrating the agro sector into the market economy.

Another Swiss colony was established in 1823 in Chabag, Bessarabia, near the mouth of the river Dniepr at the Black Sea. These were mainly immigrants from Waadland in West Switzerland (Bühler et al. 1985) and they were primarily active in the cultivation of grapes and production of wine making, bringing an early contribution in transferring wine production technology to the region.

In parallel to the increasing flow of immigrants the birth, development and consolidation of Russian bureaucracy continued. Incentives to bring foreign capital and immigration started decreasing and regulations increasing. In 1871, the Russian government repealed the manifestos of Catherine II and the special treatment of German settlers. Russian authorities saw too many poor farmers entering the country, brining little benefit to the state. Later, the Soviet regime that lasted for about 70 years after the revolution of 1917 had in any case isolated the country from any Western investments of that kind.

After the political and economic reforms in the early 1990s history started repeating itself. The change from a centrally planned to a market economy has encouraged many Western European investors to go to Russia and they were welcomed by the authorities who were eager to upgrade Russia's agriculture and food-processing technology. Apart from large industrial food corporations numerous German, Swiss, Austrian and other European farmers have given their farms and gone into farming and food production in Russia. Such investments include all kinds of farming and food products such as grains, vegetable, processed meat products, wine, dairy products and even fish. The attracting features for the investors are the same as in Catherine's time: huge local food markets, excellent fertile soil and large pieces of land that can be cultivated economically.

15.2.3 KEY SUCCESS FACTORS

The case of Catherine II and her successors illustrates the following key factors for successful technology transfer:

- The agricultural sector was probably the most important industry at that time. The government therefore followed a sound strategy when attracting foreign immigration, investment and technology to this crucial part of the economy (see Chapter 8, 'Developing a Technology Strategy').

- The government opened up to foreign investment and immigration and showed a strong commitment in supporting foreign expertise flowing into the country. In addition, Catherine II set strong policy incentives to attract immigration and foreign technology in the agro sector (see Chapter 11, 'Technology Attraction').

- The close integration of the foreign farmers, cheese and wine producers allowed for vertical technology spillovers. This facilitated the absorption of the transferred technology into the local economy and guaranteed a sustainable productivity increase (see Chapter 12, 'Technology Absorption').

15.3 Egypt: Cotton and Textile Production Technology

15.3.1 EGYPTIAN ECONOMIC DEVELOPMENT AND THE TEXTILE INDUSTRY

The textile industry is today one of the central pillars of the Egyptian economy. In 2009, it accounted for 27 per cent of Egypt's industrial production, contributing approximately 3 per cent to total GDP and employing a million workers (TDNE 2009). At present, Egypt is a major exporter of cotton, textile material and ready made garments all over the world, representing 24 per cent of Egypt's total non-oil exports (TDNE 2009). The industry's development is associated with the high quality of its long-fibre cotton. Its origins date back about 200 years.

Muhammad Ali, an Albanian born in Kavalla (today Greece, then a part of the Ottoman Empire) was a military commander sent by the Turkish authorities to Egypt to repel the Napoleonic troops that had landed in Egypt at the beginning of the nineteenth century. With the help of the British he conquered the French and then consolidated his power as ruler of Egypt. His dynasty lasted for almost 150 years, until 1952 when king Farouk was forced out by a bloodless revolution conducted by Gamal Abd el Nasser and his fellow colonels.

Mohammad Ali ruled Egypt from 1805–1848 and can be considered the father of modern Egypt. His objective was to create a modern powerful country using the European states as his ideal. He reorganized Egyptian society, streamlining the economy, training a professional bureaucracy and building a modern military. In addition to bolstering the agricultural sector he started building an industrial base for Egypt. One of his visions was to establish a textile industry in an effort to compete with European industries and produce greater revenues for Egypt.

Egypt already had a fine textile production culture that dated back to the Pharaohs. However, Mohammad Ali soon realized that in order to implement his objective and build a strong and modern textile industry, he would need to attract foreign capital and technology. To attract this FDI, he provided special incentives (see Chapter 11):

- free land for settlement;

- tax exceptions; and

- a variety of government guarantees.

In this respect he was an early and important foreign investment promoter.

15.3.2 INVESTMENTS INTO THE EGYPTIAN COTTON INDUSTRY

Among the first foreign investors were the Greek Tossizza brothers, friends of Mohammad Ali whom he had known when he was a tobacco merchant in Kavalla. The brothers invested in agricultural production and ginning of cotton, but their main strength and contribution to the economy was exporting the cotton to world markets. Many other Greek investors and immigrants followed the Tossitza brothers. In the second half of the nineteenth and first half of the twentieth centuries the Greek cotton traders played a very important role in the industrialization and particularly in the exportation of Egyptian cotton.

The Egyptian textile tradition continued over the centuries. In the seventeenth and eighteenth centuries textile production constituted the leading artisanal activity in Egypt, with a centre in the Islamic city of Cairo. A favourable business condition for Egypt was created by the American Civil War (1861–1865). During these years, the cotton-producing southern US states were cut off from world markets giving the Egyptian cotton producers the opportunity to establish themselves internationally.

However, new times and new markets demanded new *technologies*. England was particularly active in the introduction of textile technology to Egypt. As early as the 1820s, steam engines imported from England were installed at workshops in Cairo and in the Delta town of Mansura to drive flax mills (Beinin 2010). Jennies operated by animal power were modernized by employing steam engines from England. Some of the machinery for the textile production enterprises was built in Egypt with imported tools under the supervision of the Frenchman who developed long-staple cotton, F. Jumel. Another example of foreign investment was a Swiss man, a Mr Reinhart who founded Reinhart & Company in Alexandria in 1907, becoming strongly involved in Egyptian cotton. The company soon grew into one of the largest Egyptian cotton exporters and ginning industrialists. In summary, investments and technology transfer from England, France and Switzerland played a significant role in modernizing the entire textile industry including ginning, spinning, weaving, dying and garment-making (Beinin 2010).

15.3.3 KEY SUCCESS FACTORS

Although the technology transfer to the Egyptian cotton and textile industry is a historical example, the case shows the importance of several key success factors:

- With the cotton and textile industry, Muhammad Ali and his successors focused on an industry sector in which Egypt already had a prior technology base (see Section 4.2). Egypt was already in possession of fine textile production and had been since the reign of the Pharaohs. This allowed Egypt to quickly absorb new textile and cotton technologies and to upgrade its technological capabilities in this industry.

- With the selection of the cotton and textile industry as a focus sector for FDI and technology transfer, Egypt followed a sound technology strategy, reflecting two key factors. First, it exploited an external opportunity created by a growing world demand for textiles which was enhanced by the temporary loss of a major competitor during the American Civil War. Secondly, Egypt leveraged on its internal strengths, namely its favourable climate and soil, the availability of water from the River Nile, its prior technology base in textiles and its strong position for seaborne exports to Europe. By doing so, Egypt followed a robust SO strategy, as described in Section 8.4.

- Finally, Muhammad Ali set strong policy incentives to attract FDI and technology. Most importantly, this communicated the government's clear commitment to support foreign investors and to guarantee their rights (see Chapter 11).

15.4 Chile: Wine Production Technology

15.4.1 THE CHILEAN WINE INDUSTRY

The development of the Chilean wine industry over the last three decades is a typical example of how a country can add considerable value to its natural resources through foreign investment and technology, making a substantial contribution to its economic progress. Chile has produced wine for many centuries, but it is only due to technology transfer in the recent decades that its wines have attained such a high quality and become an appreciated export product all over the world.

Chile's wine history goes back to the time of Spanish colonization. As the Spanish Conquistadors and missionaries came to Chile around 1554, they brought grapevines with them. Over the centuries, unexceptional wine was produced and consumed mainly by the local population. During Spanish rule, vineyards were restricted in production of wine to ensure that Chileans would purchase wine from Spain. After Spanish rule an unfavourable political climate and high taxes over a long period of time discouraged foreign investors from entering the country and bringing the necessary expertise.

Chile has some of the best conditions for growing wine in the world. The country is a long and narrow strip extending over 4,000 km from the freezing Antarctic in the south to a hot and dry Atacama desert in the north. It is squeezed between the Pacific Ocean in the west and the Andes mountains in the east. The large variety of climates, the rich soils and the availability of water from the Andes combine to create excellent conditions for viniculture. Grapes are among the most important products of the rich agricultural variety of Chile. Further, Chile's natural boundaries (sea, mountains, desert and ice) have left the country relatively isolated from other parts of the world and been beneficial in keeping the phylloxera louse away from its grapes (MacNeil 2001: 836–843). This reduces the need to use pesticides, and some people believe that this 'purity' of the vines is a positive element that can be tasted in the wine.

15.4.2 ATTRACTION OF FDI AND TECHNOLOGY TO THE WINE INDUSTRY

It was not until the 1980s that a major shift occurred in the wine industry. The democratization of Chile in the last two decades of the twentieth century had an important effect in attracting foreign investments. The new democratic governments strongly supported foreign investment and therefore set up many tools and incentives to facilitate foreign investors enter the Chilean market. The Chilean Economic Development Agency (CORFO) created the InvestChile programme to provide local and foreign investors with special incentives and services. Financial incentives were available to investors through all development stages, specifically in relation to the provision of information and advisory services, support in searching and developing human resources, lease of real estate at attractive conditions, administrative and licensing support, etc. The newly attractive investment environment and the fact that Chile had one of the lowest rates of corruption in Latin America have played a very important role in attracting foreign investors in different sectors including the wine industry (see Chapter 11).

The development of the Chilean wines in the 1990s and early years of the twenty-first century is perhaps the most precocious growth spurt in wine-making history. Chile went from producing wines that were nothing of note to wines that were first class in less than ten years. Vineyard firms in Spain, Italy, France and the United States have invested heavily resulting in a large number of Chilean vineyards gaining some of the most up-to-date facilities and expertise in the world. This, together with the perfect wine-growing climate and cost-competitive labour force (particularly as compared to Europe) has made Chilean wine some of the most competitive in the world in terms of quality–price balance.

In particular Chile has benefited from the technology associated with foreign investment. Typical improvements were the better fermentation process in stainless steel tanks and the use of oak barrels for wine storage and maturing. Investors opened their own wineries or collaborated with existing Chilean wineries to produce new brands. This collaboration combined with the clustering of vineries in the Central Valley and its subregions (Maipo, Rapel, Curicó and Maule), provided opportunities for substantial technological spillovers to local wine producers (see Sections 6.3 and 12.2.1 on clustering). In addition to modern wine-making processes, investors developed sound marketing concepts for bringing the new improved wines to the world markets (see Chapter 13, 'Technology Application').

Due to these foreign investments and the associated technology transfer, Chile is today known as a world-class producer of cabernet sauvignon, Bordeaux, merlot and chardonnay. It is a major exporter of wine all over the world. The country is the fourth-leading exporter, after France, Italy and Australia, into the United States (MacNeil 2001: 836–843).

15.4.3 KEY SUCCESS FACTORS

The case of Chile's wine industry serves as an example to illustrate the importance of a variety of factors for successful technology transfer via FDI:

- First, the country leveraged on its internal strengths such as the large variety of climates, its rich soils, the availability of water from the Andes, its natural boundaries keeping away the phylloxera louse and its cost-competitive labour force (see Chapter 8). As a result, its firms developed a differentiation advantage through the production of high-quality wines (see Section 3.1).

- Secondly, Chile created an attractive environment for FDI. Besides other factors such as political stability and low levels of corruption, it set up the InvestChile programme to provide investors with special incentives and services (see Chapter 11).

- In addition, the geographic concentration of the wine industry allowed for the emergence of an industry cluster (see Section 12.2.1). This cluster of wine producers facilitated the spillover of technology from foreign investors to local firms.

- Finally, the investors placed emphasis on marketing the high-quality wine externally in the world market. This guaranteed that the newly developed product technology was actually applied and commercialized (see Chapter 13).

15.5 Malaysia: Automotive Manufacturing Technology

15.5.1 MALAYSIA'S ECONOMIC DEVELOPMENT

Malaysia is another representative example of how a country can achieve substantial technological, industrial and economic progress through the attraction of foreign technology within a timespan of only a few decades (see also Malairaja and Zawdie 2004: 233). Since 1980, the Malaysian economy has grown at an historically unprecedented average rate of 6.1 per cent a year (IMF 2009, Figure 15.1). This strong growth has resulted in real GDP increasing by a factor of more than five times over the same time period. The volume of highly technological manufactured exports, notably electronic products, increased rapidly. The country has achieved the status of a newly industrialized country. Malaysia is therefore considered one of the most successful non-Western countries to have achieved a relatively smooth transition to industrialization and rapid economic development. We will now discuss some of the reasons why Malaysia was able to achieve such an economic success.

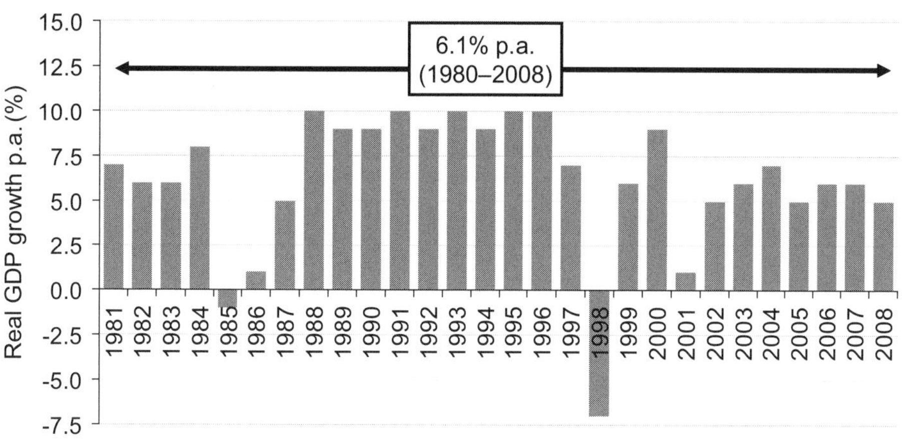

Figure 15.1   Malaysian real GDP growth
Data source: IMF.

The Federation of Malaysia was formed in 1963 from Malaya (now known as Peninsular Malaysia), Singapore, and Sarawak and Sabah which are the two territories known as East Malaysia on the island of Borneo (due to internal political tensions Singapore left the Federation in 1965). Prior to the establishment of the Federation, the territories had a very small technological basis. Existing industry was related to processing and exporting primary products such a tin, rubber, palm oil and wood. There was also a limited production of manufactured products for the local market (building materials, biscuits, cigarettes, beverages, etc.) (Drabble 2000).

In the late 1960s the country faced ethnic conflicts, in particular between the poorer Malays, who formed the majority of the population of the country, and the economically better off Chinese who controlled a substantial part of the country's trade. The government realised that the best way to maintain ethic harmony and peace was to make an intensive effort to eradicate poverty irrespective of ethnic background. The main consideration in achieving this aim was that the redistribution should not result in any one group losing in absolute terms. Rather it should be achieved through the process of economic growth, that is more investments and know-how had to be attracted and more jobs created. In this respect Malaysia is probably one of the best example of a country in which the economic roles and interests of various ethnic groups have been pragmatically managed in the long term. Despite the initial conflicts a harmonic symbiosis between the different groups has been maintained (Drabble 2000).

Starting from the early 1970s there has been a rapid development towards a range of export-oriented manufacturing industries such as textiles, electrical and electronic goods, rubber products, etc. (Ismail 1999). While the primary sector would continue to receive developmental aid under the successive Five-Year Plans, the main emphasis was a switch to export-oriented industrialization with Malaysia seeking a share in global markets for manufactured goods. Free Trade Zones were set up in places such as Penang where production was carried out with the undertaking that the output would be exported (Ismael 1999: 22, see Section 12.2.1). Firms locating there received concessions such as duty-free imports of raw materials and capital goods as well as tax concessions, aimed at primarily foreign investors who were also attracted by Malaysia's good facilities, relatively low wages and docile trade unions. A range of industries grew up: textiles, rubber and food products, chemicals, telecommunications equipment, electrical and electronic machinery/appliances, car assembly and some heavy industries, iron and steel. Much of the capital and technology was foreign. Malaysia owes this successful industrial and economic development to a number of factors.

- *Geographically* it lies at a crossroads of important trade routes bringing to the country an exposure to the global economy.

- *Government policies* have instrumental in creating an environment attractive to foreign investors and thus pulling in the required capital and particularly technology needed to industrialize the country.

- The rapid industrialization and growth in production for export was facilitated by an accompanying development of the country's *infrastructure* including roads, railways, ports and a dense and efficient telecommunications system. Well-maintained roads linked all parts of the country. This was complemented by railways which traverse the west and east coasts of Peninsular Malaysia. Today Malaysia has eight international airports and six major ports.

15.5.2 MALAYSIA'S AUTOMOTIVE INDUSTRY

Automotive technology transfer to Malaysia, originally from Europe but then more intensively from Japan, has made a considerable contribution to the development of the country. The special benefit of the automotive sector is that it is a large cluster with long and wide supply chains (mechanical, electrical and electronic, synthetic materials, etc.) and therefore has a strong impact on enhancing the overall technological basis of industry, increasing absorptive capacity across various industries (see Section 4.2). The automotive industry in Malaysia can therefore be considered as one of the most important and strategic industries in the manufacturing sector. Compared with others, the sector has been earmarked to boost the industrialization process so that Malaysia can be a developed nation by 2020 (UN ESCAP 2002: 71–83). Malysia's automotive industry went through several phases (UN ESCAP 2002: 71–83):

- In 1967, the government approved the operation of the first car assembly plants. First were some European (Volvo and Peugeot) and Japanese (Mazda and Nissan/Datsun) companies who started assembling their cars in Malaysia. At that time the level of technology transfer was still low as was the development of human resources in the industry. Also, at the beginning of the 1980s the Malaysian automotive industry suffered from structural problems such as inefficient assembly plants, small volumes and a wide variety of models and brands (Leutert and Südhoff 1999: 251).

- The rapid development of the Malaysian automotive industry began with the launch of the National Car Project, *Proton*, in 1984. The Proton project was a joint venture programme of Mitsubishi Motors Corporation of Japan and a local partner, the state-owned Heavy Industries Corporation of Malaysia (HICOM) (see Malairaja and Zawdie 2004: 242). Government support and intervention for this project was strong (Leutert and Südhoff 1999: 250). In the beginning the joint venture only held assembly capabilities and most of the components were imported, but over time, manufacturing capabilities were developed (see Section 2.5) and there was a gradual shift to almost complete local integrated production. As an integrated manufacturing project, it was given preferential tax and duty rates (see Chapter 11).

- After the success of the first national car, the Perodua car project was established in October 1992. It was the result of an agreement between UMW Corporation, Daihatsu Motor Company Ltd of Japan, Med-Bumikar Mara, PNB Equity Resources Corporation, Mitsui and Company Ltd of Japan and Daihatsu (Malaysia). Perodua was set up to expand the automotive product range and to further support components and parts manufacturing.

The Malaysian auto market is today dominated by Malaysia's national cars, Proton and Perodua. In 1999, Proton was the number one brand of car not only in Malaysia, where it commands a market share of roughly 70 per cent (UN ESCAP 2002: 76). The automotive industry in Malaysia today consists of 4 car manufacturers, 15 car assemblers, 3 composite body/sports car makers and 350 part suppliers producing a wide range of parts with a substantial technological know-how (safety glass, castings, pistons and piston liners, transmission and steering parts such as a gear shift components, drive shafts, clutches, shock absorbers, brake drum, brake disk and brake parts, batteries, voltage regulators, electronic controls, etc.) (UN ESCAP 2002: 8).

Nevertheless, today the Malaysian automotive industry faces increasing competition from lower-wage countries, especially India and China. The industry will have to strengthen its competitiveness through greater emphasis on product and market development and the acquisition of R&D capabilities (see Section 2.5), allowing it to develop a differentiation advantage (see Section 3.1). In addition, the automotive industry faces the challenge of a scarcity of technically skilled labour (Leutert and Südhoff 1999: 256). Finally, it is noteworthy that the contribution of joint venture activities and FDI flows to the development of innovation capability in Malaysia has been less successful in other industry sectors (see Malairaja and Zawdie 2004: 243).

15.5.3 KEY SUCCESS FACTORS

We have already mentioned some key success factors for Malaysia's economic development, such as the development of infrastructure and its government policies. For the transfer of technology to the automotive industry, two factors were favourable:

- For the automotive industry, Malaysia focused on an industry with long and wide supply chains. This allowed the country to develop an extensive industrial base and facilitated technological spillovers to a variety of other industries, broadly increasing the country's absorptive capacity (see Section 4.2).

- In the case of the Proton project, the formation of a joint venture with Mitsubishi Motors and strong governmental support facilitated the establishment of a national automotive industry and the move from assembling to manufacturing capabilities (see Section 12.2.2). However, the joint venture also resulted in a limited autonomy of HICOM vis-à-vis Mitsubishi (Leutert and Südhoff 1999: 251). In addition, it is noteworthy that the strong government intervention, for example via import tariffs on foreign cars, had negative effects such as a lack of competition resulting in an obstacle to further upgrade Proton's technological capabilities (see Leutert and Südhoff 1999).

15.6 Estonia: Banking Technology

15.6.1 ESTONIA'S SUCCESSFUL ECONOMIC TRANSFORMATION

Estonia emerged as an independent nation after the disintegration of the Soviet Union in the early 1990s. Like most of the Eastern European states, this small Baltic Sea country had a difficult start when changing its economic and political system at a rapid pace. The end of the socialist regime created chaos in the country: shops were empty, the Soviet rouble had no real value anymore, industrial production declined by more than 30 per cent in 1992, wages fell by 45 per cent and overall inflation was running at more than 1,000 per cent (Laar 2007). The new political and business leadership consisted of young people who had no experience of market economy mechanisms. However, this lack of experience was more than compensated by their enthusiasm, creativity and hard work.

The pace of change was amazing. Within about 15 years Estonia was transformed from a communist Soviet state to a country with an economy amongst the freest in the world. Estonia became the first former communist country to rise to the status of a 'free' economy in the annual Index of Economic Freedom, produced by The Heritage Foundation and *The Wall Street Journal* (WSJ 2009). In 2009 it ranked sixteenth in the Index of Economic Freedom, ahead of traditionally free economies such Finland, Sweden, Germany and Austria (WSJ 2009).

The success of the transformation is attributed to a number of factors, but in summary it can be said that most important have been:

- the speedy and wise reforms undertaken by the government; and

- the substantial FDI encouraged by these reforms.

After the first democratic elections since the Second World War took place in September 1992, the new government, consisting of mostly young ministers in their early thirties, realized that the best way to proceed was to act very quickly and reform the country from its roots. Their assumption was that a radical economic programme launched as quickly as possible has a much greater chance of being accepted than either a delayed radical programme or a non-radical alternative that introduced difficult measures gradually. Important reforms (see Laar 2007 and our discussion of policy measures for attracting FDI in Section 11.1) were:

- In 1992, Estonia became the first country of the former Soviet Union to introduce its own currency. Using a currency board system, the Estonian Kroon was made fully convertible from the first day by *pegging it to the German mark*. Fixing the exchange rate to a strong currency like the Deutsche Mark created the necessary trust for a new wave of foreign investors to the country. The key objectives of the reform were eliminating inflationary impacts from the east, guaranteeing an equilibrium exchange rate based on supply and demand and conquering the cash crisis.

- The government also realized that monetary reform cannot succeed unless the budget is strictly controlled. *Balancing the budget* required radical cuts in all kinds of subsidies for state-owned companies. It also implied reducing the size of government. These cuts were unpopular and not very easy to implement but the government managed to push them through, thanks to the coalition agreement between the parties which established balancing the budget as the most important goal. Later a law was passed that only a balanced budget can be presented to the Estonian parliament. Since then a balanced budget has become a standard for which Estonia is well known.

- Estonia *reduced trade tariffs and non-tariff barriers and abolished all export restrictions*, making the nation a free-trade zone. This reoriented the economy from the East to the West, and exports began to grow rapidly.

- Estonia did *not* try to encourage foreign investors by offering all sorts of enticements, such as tax exemptions, privileges and special rights (see also our discussion of FDI incentives in Section 11.2.1).

Instead, it worked to *create a business environment that favours both local domestic investment and foreign investment* without making any distinction between them. Passage of the law on the sale of land ensured that all foreign investors could feel a greater sense of security, and also signalled that their property rights would be protected. This strategy worked very well. Between 1993–1994 Estonia went from an almost unknown location for foreign investors to a preferred investment destination in the region. This large inflow of investment created new working places, reconstructed old factories and brought new knowledge and technology.

- The new Estonian government understood that uncertain or non-transparent rules, heavy regulation and pervasive controls give officials exceptional power, many opportunities to seek bribes, and a wide scope for appropriating public wealth. In this respect, the government understood that the most effective method of dealing with *corruption* and organized crime is decisive implementation of market economy reforms, the development of a civil society and the rule of law. Thus one of the priorities of the government was to make rules transparent and clear, reduce controls on foreign trade, remove entry barriers to private industry and privatize state companies. Eliminating subsidies, 'soft' loans and all other such privileges removed another inducement for bribes.

- At the heart of economic transition was a shift to private ownership. This included restitution of property to former owners, *privatization of existing state assets, establishment of laws and regulations as well as enforcement mechanisms that secure the property rights of individual and legal persons. This was one of the decisive measures that encouraged foreign investment.

- One of the most radical changes of the government was the introduction of a *flat tax rate* (instead of a progressive tax rate). The basic philosophy was to encourage entrepreneurship by not punishing success. The government was convinced that the entire tax system should favour savings and investments and encourage people to create new wealth. The tax system in Estonia had to be simple, inexpensive to apply, transparent and understandable to the taxpayers. The tax base should be as broad as possible with a minimum number of exemptions, minimizing incentives for tax avoidance such as the underground economy. The tax rates had to be low, encouraging the economic activity of people and creating more growth.

- The government very quickly appreciated that good laws by themselves are not enough. It acted fast to build *effective institutions* that moved the new laws from paper to practice. A new system of judges, prosecutors, arbitrators, court functionaries and enforcers of law was established and a private legal profession was developed.

15.6.2 TECHNOLOGY TRANSFER TO ESTONIAN BANKING

Amongst the most important foreign investments were those made by Scandinavian banks. These investments carried the missing and much needed technology to run the newly formed market economy of the country.

During the Soviet era the primary function of the banks was to control the distribution of funds to state enterprises according to a central planning scheme laid down by the government. Granting credits to enterprises on the basis of business viability analysis was not practised at all. The banks had almost no expertise in credit management, business plan assessment or treasury management, all key elements of a financial system. The hard technology (see Section 2.3.1) applied by the banks, including the use of IT software and hardware, was antiquated and very inefficient. The government had quickly realised that acquiring expertise in the banking sector was crucial and urgent if the transformation to a market-driven economy was to succeed. It understood that choosing to finance only viable businesses was a key success factor to a market economy.

In the early 1990s, multilateral development institutions such as the World Bank or the European Bank for Reconstruction and Development (EBRD) were offering development aid and technical assistance to Eastern European countries and countries that had emerged from the implosion of the Soviet Union. Many West European and North American donor governments did the same. Institutions were sending hordes of consultants to advise and train the banks. This development aid was most welcome by almost all countries in the East. However, Estonia adopted and entirely different philosophy. The country rejected much of the offered development aid and chose to acquire the necessary explicit and tacit technology much faster and much more effectively by opening the banking sector to foreign investors.

The much needed capital injection, banking know-how and technology came from two major Swedish banks, Skandinaviska Enskilda Banken (SEB) and Swedbank. The strategic acquisitions by these two banks have led to the emergence of two large competing banking groups in the Baltic countries. Swedbank was the first to reach the Baltic market, by acquiring a strategic stake in Estonian Hansabank which held a dominant position in the Estonian market. Soon after this SEB bought stakes in Eesti Ühispank in Estonia, Latvian Unibanka in Latvia and Vilniaus Bankas in Lithuania. Through these investments the Estonian banking sector was fully integrated into the Scandinavian banking circles by the year 2001 (Estonica 2009). Scandinavian banking groups owned stakes in four major commercial banks that together covered 97 per cent of certain parts of the market (Estonica 2009). Investment by Swedbank and SEB has improved the banks' credibility considerably and stabilized their balance sheets.

Most importantly, the Scandinavian banks have been instrumental in *transferring the necessary banking technology* that helped increase the profitability of the banks and professionally finance and stimulate the overall economy. As part of this process, Estonian banks were active in employing new *hard technology*: by the end of 2000, the four major banks boasted 265,000 Internet and 139,000 Telebank customers. Such indicators put Estonia at the same level with most Western European countries regarding the number of electronic bank users in the course of a few years. In addition, *soft technology* (see Section 2.3.1) was transferred by sending expatriates into key management positions of the acquired banks, by introducing new business processes and standards (for example for accounting), and by training many local personnel (see Section 6.1).

15.6.3 KEY SUCCESS FACTORS

The successful economic transformation of Estonia is tied to a variety of factors. Successful technology transfer to the banking industry is one of them and was facilitated by a variety of drivers:

- Similar to Malaysia, Estonia focused on a key industry sector to attract FDI and foreign technology. Through credit markets and financing, the banking sector had important interfaces to all other industries, facilitating the spillover of productivity improvements throughout the entire economy (see also Chapter 8).

- However, these banking investments would never have occurred unless the government had not undertaken major reforms to create an environment attractive to foreign investors (see Section 11.1). Commercial branches of banks (not investment banks), and particularly West European banks, are in general relatively conservative institutions. They have no an interest in becoming deeply engaged in economies that are not well structured and regulated and in which the prospects of economic development are not present. Radical reforms of the government were therefore a critical precondition for the market entrance of the banks.

- Finally, the transfer of banking technology to Estonia is an example of a successful internal transfer between MNCs and their local subsidiaries (see Section 6.1). A major driver for this is the value chain of the banking industry. In manufacturing the various phases of value creation can often be geographically split apart, allowing MNCs to withhold certain technologies. However, banking is a service industry. Large parts of value creation are achieved locally. This required the Scandinavian MNCs to transfer technology to their local subsidiaries in order to increase the productivity of their local banking activities.

15.7 Ireland: ICT and Software Development Technology

15.7.1 IRELAND'S ECONOMIC DEVELOPMENT

When Ireland joined the European Union (EU) in 1973, it was labelled 'the poorest of the rich' because its national income was a mere 64 per cent of the EU average (Sands 2005: 42). In the years that followed Ireland developed into a rich country with a GDP per capita of USD 40,000 (2009): one of the highest in the world that year (IMF 2009).

One of the key drivers for this rapid development has been the ability of the country to attract a large amount of FDI and particularly to attract businesses with high technology and added value content. Substantial foreign investments have been made to the pharmaceutical, electronics and software development sectors, the last being of significant importance because of its particularly high added value and its contribution in elevating the overall technological level of the country, increasing Ireland's general absorptive capacity (see Section 4.2). Figure 15.2 indicates that this trend accelerated rapidly at the beginning of the 1990s, reaching a peak in 2002 with a FDI inflow of close to US$30 billion per year.

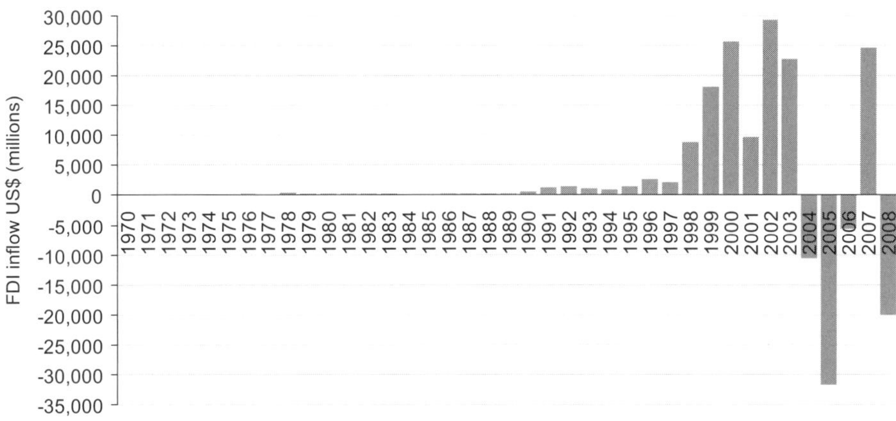

Figure 15.2   FDI inflows to Ireland in US$ million
Data source: UNCTAD.

The bulk of the investments with technological content came from the USA. Particular advantages of Ireland for American MNCs were:

- A highly educated workforce, as Ireland ranks amongst the best educational systems in the world (see Section 12.1.1).

- English speaking (see Section 5.4.2.3).

- For most of the 1980s and 1990s Ireland had some of the cheapest labour in the EU (see Section 5.4.2.2). This has been a particularly important consideration for foreign investors because of the labour intensity of the software and electronics industries.

- Historical and cultural ties to the USA as many Americans in high political and business positions are of Irish descent (see Section 5.4.2.3).

- Proximity of Ireland to Europe and accessibility to the large European Union markets (see Section 5.4.2.2).

- Ireland is a beautiful country with old traditions and culture and with a well-established infrastructure which makes it attractive for expatriates of foreign companies to relocate there (see Section 5.4.2.2).

Despite all the above inherent features it would have not been possible to attract so much FDI unless the government opened the economy and created regulations and incentives attractive to foreign investors. Throughout the 1970s, 1980s and 1990s the government maintained its commitment to facilitating investment processes. Among the most important measures were the following (Drew and Foster 1994):

- Ireland's *wages* remained low mainly due to partnership agreements between workers and the government, which were established as part of the Programme for National Recovery in 1987. Through these agreements, workers accepted lower wage increases in return for tax cuts that provided a significant increase in actual take-home pay.

- With the increasing flow of FDI, the country started facing acute labour supply shortages, which also threatened the rise of labour costs. In addressing the labour shortage the government started a massive programme of *attracting skilled immigrants* to Ireland (see Section 12.1.1.2: Attracting human resources). The state training and employment agency held job fairs all over Europe and North America. In the late 1980s and 1990s, about a quarter of a million skilled people joined the Irish workforce from abroad.

- The Industrial Development Agency (IDA) Ireland, which is the national organization responsible for attracting foreign investment, was given adequate funding for projects that targeted the development of high-tech industries, particularly focusing on attracting investment from the US.

- In 1981, the government introduced a low tax rate of 10 per cent for international manufacturing companies. Other corporate taxes in the country were as high as 25 per cent, but few international companies paid this rate as there was some degree of flexibility on how 'manufacturing' was defined. The dual corporate tax structure of the country has led to complaints of unfair tax favouritism by other EU countries and Irish companies that did not qualify for tax deductions. The government responded with a new corporation rate of 12.5 per cent that was applied to all business sectors starting form 1 January 2003. Ireland established one of the lowest rates in the EU.

- In addition to a low tax rate, Ireland has also developed *non-financial incentives* for businesses. The key measure was to reduce bureaucracy, speed up processes and reduce the administrational effort required by a foreign investor to set up in the country. In an action programme of regulatory reform the government has placed responsibility on each department or office to revise, consolidate, or repeal frivolous laws within their jurisdiction, especially those that hinder market entry or are burdens to small businesses. In addition, the tax authorities have made an effort to reduce the time and effort required for tax payments. This has been done by consolidating taxes, creating electronic payment options and maintaining high levels of customer service.

These measures began encouraging investments to Ireland from high-tech US corporations that were increasingly penetrating global markets in the 1970s–1990s. The objective of most of these investments was to use Ireland as a base for exporting to Europe; eventually transforming the country into an export-driven economy.

15.7.2 FDI AND TECHNOLOGY TRANSFER TO IRELAND'S SOFTWARE INDUSTRY

One sector of the Irish economy in which FDI and technology transfer played an important role was Information and Communication technology (ICT) and in particular the software industry. The Irish software industry evolved from humble beginnings comprising just 365 firms and employing fewer that 8,000 persons in 1991. By 2003 there were already 900 software companies, employing close to 24,000 persons (Sands 2005: 45). In recent years, Ireland has become one of Europe's top locations for software development. Today, the ICT sector in Ireland has more than 300 overseas companies, employing approximately 60,000 people out of a total of 90,000 in the entire ICT sector (Cassidy et al. 2009: 58). More critically, the sector creates a total revenue of approximately €50 billion (IDA 2009). Seven of the top ten leading ICT companies have a substantial base in Ireland (IDA 2009). Among the most important software and hardware computer companies that invested in Ireland are Apple Computers, Compaq, Computer Associates, DEC, Dell, Hewlett-Packard, IBM, Intel, Microsoft, Oracle and Sun Microsystems. Also significant were major US banks, airlines and other industrial corporations whose software was fully develop in Ireland. In overall, in 20 years Ireland has accumulated a tremendous amount of know-how in all software development areas:

- Application of computer software to industrial processes, for example automation in sectors such as automobile, chemicals, textiles and engineering manufacture.

- Implementation of digital techniques in telecommunications.

- Development of office automation, to support functions ranging from word processing to highly complex distributed processing systems evident in the banking and insurance environment, airline booking systems, accounting systems, invoicing systems, etc.

- Development of software relating mainly to the 'leisure' industry (video games, videotext services, etc.).

Besides the foreign-owned software sector, Ireland has also developed a sizeable indigenous sector. The emergence of this sector occurred largely in parallel to the establishment of MNCs in Ireland. Indigenous companies are characterized by a strong niche focus with small volumes and an export orientation (Cassidy et al. 2009: 58). There are several indicators that the indigenous sector enjoyed technological spillovers from MNCs (see Chapter 6). For example, companies in the indigenous Irish software industry have often said that they have been helped to developed new or improved software products by their interaction with foreign-owned MNCs (O'Malley et al. 2008: 169). Also, in terms of work experience, around one-third of Ireland's indigenous software entrepreneurs were employed at MNCs before starting a company of their own (Cassidy et al 2009: 60). In addition, some software companies emerged as spin-offs of MNCs, such as Airtel ATN (from Verel, a US telecommunications company) or Allfinanz (from Siemens) (Sands 2005: 50).

15.7.3 KEY SUCCESS FACTORS

We have already identified a variety of factors such as language and the proximity to Europe as critical factors for the success of Ireland in attracting FDI.

Beyond these, the following factors were of particular importance for technology transfer to the ICT sector:

- The government played a very active and aggressive role in attracting FDI to the ICT sector. A low tax rate, the effectiveness of the Industrial Development Agency (IDA) Ireland and the government's professionalism in implementing its foreign investment policies supported the movement of major MNCs to Ireland.

- For the ICT sector, the availability of relatively cheap but at the same time well-qualified human resources was of crucial importance. Ireland has created an education system that is specifically designed to meet the needs of industry, providing an ample supply of highly skilled graduates. For example, Ireland invested 44 per cent of its structural funds received by the EU from 1994–1999 to human resources and only 20 per cent to physical infrastructure (compared to Greece, which distributed only 25 per cent of its funds to human capital and 40 per cent to infrastructure). Also, the government significantly restructured its higher educational institutions to meet the needs of industry. These reforms proved even more effective due to the fact that in 1999, 47 per cent of the Irish population was aged 29 and under, providing a large supply of young labour moving through the education system (see Sands 2005: 60–62). As a result, in the Global Competitiveness Report, the country ranks eighth in the world regarding the quality of its educational system (WEF 2009: 396). The availability of qualified labour was additionally supported by a growing number of immigrants (see Section 12.1.1.2: Attracting human resources). Most interestingly, large parts of these immigrants were returning Irish emigrants who had gained working experience aboard (Sands 2005: 66).

- The presence of MNCs in the Irish software sector facilitated the emergence of an indigenous Irish software industry. This allowed for spillovers of technology to local firms through mechanisms such as spin-offs, labour mobility and supplier–customer linkages. An important factor in this respect was that domestic firms did not compete with MNCs but rather focused on niche markets with highly specialized products. This reduced barriers to transfer technology from the MNCs to the indigenous software firms.

15.7.4 THE ECONOMIC DECLINE OF 2010

The Irish economy expanded rapidly during the 'Celtic Tiger' years from the 1980s to 2007. However, between 2008–2010 the country experienced a major financial crisis that was responsible for Ireland falling into recession for the first time since the 1980s. The Irish government officially announced it was in recession in September 2008, with a sharp rise in unemployment occurring in the following months.

Ironically, many of the factors that have contributed to the growth of the economy over the last 25 years or so were proven not to be sustainable: the economy of the country was too weak to resist the global financial crisis of 2008–2010. Important factors that have contributed to the decline are:

- One important incentive that attracted FDI was the low labour costs of Ireland. However, Eastern European countries with equally competitive or even lower labour costs and skilled labour have joined the European Union since then. Furthermore, apart from low production costs these countries, such as Poland, offered a larger market potential. The policy of reliance on low-cost labour proved not to be economically sustainable in attracting FDI in the long term.

- In addition to the declining FDI and economic activity the country had to provide benefits to its unemployed citizens and also to the many unemployed immigrants. The numbers of people claiming unemployment benefit in Ireland rose to 326,000 in January 2009, the highest monthly level since records began in 1967.

- Low corporate taxation has deprived the government of adequate financial resources and reserves to deal appropriately with the crisis.

- The euphoria of the strong economy growth years led to an expansion of credit and included a property bubble which burst in 2007–2008. This left the Irish banks highly exposed.

As in the case of Greece, Ireland had to turn in 2010 to the European Union for economic assistance in preventing a sovereign bankruptcy of the country.

16

Summary and Conclusions

In this book, we have explained the theory and supplied practical illustratations of the importance of technology for the economic growth and competitive advantage of nations. Few if any countries have succeeded in achieving and sustaining high growth levels without investing in and attracting new technologies.

As the gap in technological advancement between industrialized and developing countries continuous to be large, the transfer of technology to the latter remains of crucial importance. FDI is one of the major channels through which this transfer can take place.

The importance of technology transfer via FDI implies that governments cannot simply wait until new technologies arrive in their domestic domain. Instead, they have to systematically manage the process in identifying, attracting, absorbing and applying suitable technologies.

This is even more the case in the light of the economic crises which struck the global economy in 2008. A reduced access to finance, a negative market outlook and a growing risk aversion have lead to a sharp drop in global FDI flows. These flows are likely to increase again in the future, but the economic crisis has resulted in an intensified global competition for scarce investments and technologies. In addition, companies in emerging markets, investment promotion agencies and governmental bodies of developing economies are becoming increasingly professional in attracting FDI and technology. This calls for a systematic management of the entire technology transfer process.

In the first part of this book (Part I), we have therefore laid out the conceptual and theoretical foundations for technology transfer via FDI. We have defined technology, described its basic characteristics and types, explained its importance for companies and nations and portrayed various theories of technology transfer (Chapters 2, 3 and 4). In addition, we have illustrated the various forms of FDI, its determining factors and contributions and the role of technology transfer in FDI (Chapters 5 and 6).

In the second part of the book (Part II), we used this conceptual understanding to provide the necessary insights into the practical process of managing technology transfer. We have shown how a technology strategy can be developed (Chapter 8), how information on technologies can be gathered (Chapter 9) and how technologies can be assessed (Chapter 10). Subsequently, we have shown how technologies can be attracted (Chapter 11), absorbed (Chapter 12), applied (Chapter 13), and how the entire process can be evaluated in terms of performance measurement (Chapter 14). In order to enhance understanding of each phase of the process, we have described a set of concepts and methods allowing for a systematic analysis and the structuring of knowledge. We closed Part II of the book by describing a set of integrated case studies and the relevance of different key success factors for each of these cases (Chapter 15).

In summary, it is difficult to state general rules for the successful attraction and absorption of technology. Rather, the entire process is highly context-specific. A country's internal factors such as its prior technology base, geographic characteristics, market size or natural resources all influence to a changing degree the absorption of new technologies. So do external factors such as international markets, competing regions and countries, technological trends or international laws. For example, for Ireland the existence of a well-developed educational system, ranking eighth in the world (WEF 2009: 396), was an important success factor in acquiring FDI and technology to the software industry. However, with Chile's education system ranking 107th (WEF 2009: 396), the same factor was not of particular relevance for the acquisition of technology to its wine industry. This is why we have focused on describing the necessary concepts and methods for analysis instead of stating general imperatives. Nevertheless, the theoretical concepts and empirical case studies illustrated in this book allow us to describe the following implications for practice:

- Governments have to develop a systematic economic strategy and, based on this, a *sound technology strategy*. This involves making the right strategic decisions in attracting investments with technologies appropriate and compatible to the available human and natural resources, economic and social characteristics as well as the prior technology base of the country (see Chapter 8). It also requires the involvement of all major stakeholders in the process of strategy formulation. On the one hand, such a move provides the necessary information regarding internal and external influencing factors on strategy formulation. On the other hand, the early involvement of stakeholders will facilitate the implementation of the technology strategy later on (see the upcoming point).

- Technology strategies have no practical relevance if they are not *actually implemented* by the respective government (see section 8.5). It is therefore of crucial importance that the government dedicates sufficient financial and human resources to carrying out the technology strategy, systematically structures the execution of measures and continually oversees their implementation.

- Governments need to continuously monitor and *update their strategies and policies*. Circumstances in markets change fast, new technological trends may occur and the logic of industries and companies continuously transforms, affecting the readiness of MNCs to transfer technology and potentially calling for a modification in policies.

- Governments have to understand that it is of importance to *open up to FDI and technology* in order to advance the socio-economic development of their country. In addition, governments need to show great commitment and become active in implementing measures for creating a business environment attractive to foreign investors. Such measures include reducing administrative barriers, accelerating processes for setting up new businesses, enhancing the protection of property rights of investors and providing a stable legal and political environment. As the case of Ireland and other countries shows, investment promotion agencies can play an important role in this context (see Chapter 11).

- This active role of governments does not mean that governments should become heavily interventionist into the economy. It is important that market mechanisms are not strangulated by trade restrictions or FDI and technology transfer incentives. Following a protectionist economic policy (for example via introduction of import tariffs or quota) in order to support the technological development of local industries may eventually result in market inefficiencies, a potential loss in competitiveness and lower economic growth. It is therefore important that governments find the right balance between the advantages of a free market economy and the policy requirements for attracting FDI and technology.

References

Adner, R., and Levinthal, D. 2001. Demand Heterogeneity and Technology Evolution: Implications for Product and Process Innovation. *Management Science, 47*: 611–28.

Agarwal, R. and Tripsas, M. 2008. Technology and Industry Evolution. In Shane, S. (ed.) *Handbook of Technology and Innovation Management.* San Francisco: Wiley: 3–55.

Aitken, H. 1985. *The Continuous Wave: Technology and American Radio, 1900–1932.* Princeton: Princeton University Press.

Akrami, F. 2008. *Foreign Direct Investment in Developing Countries: Impact on Distribution and Employment. A Historical, Theoretical and Empirical Study.* Fribourg: Faculty of Economics and Social Sciences.

Albernathy, W. 1978. *The Productivity Dilemma.* Baltimore: Johns Hopkins University Press.

Alexander, J. 1988. *Catherine the Great: Life and Legend.* New York: Oxford University Press.

Al-Laham, A. 2003. *Organisationales Wissensmanagement. Eine strategische Perspektive.* München: Verlag Vahlen.

Andrews, K. 1971. *The Concept of Corporate Strategy.* Homewood: Irwin.

APEC Center for Technology Foresight 1998. *Water Supply and Management in the APEC Region.* Bangkok: The APEC Center for Technology Foresight.

APEC Center for Technology Foresight 2002. *Nanotechnology: The Technology for the 21st Century.* Bangkok: The APEC Center for Technology Foresight.

Austin, R. and Larkey, P. 2002. The Future of Performance Measurement: Measuring Knowledge Work. In Neely, A. (ed.) Business *Performance Measurement. Theory and Practice.* Cambridge: Cambridge University Press: 321–342.

Baines, P., Fill, C. and Page. C. 2008. *Marketing.* Oxford: Oxford University Press.

Begg, D., Fischer, S. and Dornbusch, R. 1997. *Economics.* London: McGraw-Hill.

Beinin, J. 2010. Egyptian Textile Workers. From Craft Artisans Facing European Competition to Proletarians Contending with the State. In van Voss, H., Hiemstra-Kuperus, E. and van Nederveen Meerkerk, E. (eds) *The Ashgate Companion to the History of Textile Workers, 1650–2000.* Farnham: Ashgate: 171–198.

Bitzer, J. and Kerekes, M. 2008. Does Foreign Direct Investment Transfer Technology Across Borders? New Evidence. *Economics Letters, 100*: 355–358.

Blalock, G. and Gertler, P. 2005. Foreign Direct Investment and Externalities. The Case for Public Intervention. In Moran, T., Graham, M. and Blomström, M. (eds) *Does Foreign Direct Investment Promote Development?* Washington: Institute for International Economics & Centre for Global Development: 73–107.

Blomström, M, Kokko, A. and Zejan, M. 2000. *Foreign Direct Investment: Firm and Host Country Strategies.* London: Palgrave.

Blomström, M. and Kokko, A. 1998. Foreign Investment as a Vehicle for International Technology Transfer. In Navaretti, G., Dasgupta, P., Mäler, K. and Siniscalco, D. (eds) *Creation and Transfer of Knowledge. Institutions and Incentives.* Berlin: Springer: 279–311.

Blundell, R., Griffith, R. and Van Reenen, J. 1999. Market Share, Market Value and Innovation in a Panel of British Manufacturing Firms. *Review of Economic Studies, 66*: 529–554.

Bogers, M. 2009. *The Sources of Process Innovation in User Firms: an Exploration of the Antecedents and Impact of Non-R&D Innovation and Learning-by-Doing.* Lausanne: École Polytechnique Fédérale de Lausanne.

Borgatti, S. and Foster, P. 2003. The Network Paradigm in Organizational Research: A Review and Typology. *Journal of Management, 29*: 991–1013.

Branstetter, L., Fisman, R. and Foley, C. 2006. Do Stronger Intellectual Property Rights Increase International Technology Transfer? Empirical Evidence from U.S. Firm-level Panel Data. *Quarterly Journal of Economics, 121*: 321–349.

Braun, E. 1998. *Technology in Context. Technology Assessment for Managers.* London: Routledge.

Bühler, R., Gander-Wolf, H., Goehrke, C., Rauber, U., Tschudin, G. and Voegeli, J. 1985. *Schweizer im Zarenreich – Zur Geschichte der Auswanderung nach Russland.* Zürich: Verlag Hans Rohr.

Carlin, W., Fries, S., Schaffer, M. and Seabright, P. 2001. *Competition and Enterprise Performance in Transition Economies: Evidence from a Cross-Country Survey.* Working Paper 63. London: European Bank for Reconstruction and Development.

Cassidy, J., Barry, F. and van Egeraat, C. 2009. Porter's Diamond and Small Nations in the Global Economy: Ireland as a Case Study. In Bulcke, D., Verbeke, A. and Yuan, W. (eds) *Handbook on Small Nations in the Global Economy. The Contribution of Small Enterprises to National Economic Success.* Northampton: Edward Elgar Publishing: 50–134.

Catherine II 1763. *Manifesto of the Empress Catherine II on Foreign Immigration.* Issued July 22, 1763.

Chen, S. 2006. Extending Internationalization Theory: A New Perspective on International Technology Transfer and its Generalization. In Kotabe, M. (ed.) *International Marketing. Volume II*. London: Sage: 291–312.

Chenery, H. 1960. Patterns of Industrial Growth. *American Economic Review, 50*: 624–654.

Clarke, G. 2005. *Do Government Policies that Promote Competition Encourage or Discourage New Product and Process Development in Low and Middle-income Countries?* World Bank Policy Research Paper, No. 3471. Washington: World Bank.

Cohen, G. 2004. *Technology Transfer. Strategic Management in Developing Countries*. London: Sage.

Cohen, S. 2002. *Negotiating Skills for Managers*. New York: McGraw-Hill.

Cohen, W. and Levinthal, D. 1990. Absorptive Capacity: A New Perspective on Learning and Innovation. *Administrative Science Quarterly, 35*: 128–152.

Cooper, A. and Schendel, D. 1976. Strategic Response to Technological Threats. *Business Horizons, 19*: 61–69.

Dorf, R. and Byers, T. 2008. *Technology Ventures. From Idea to Enterprise*. Boston: MacGraw-Hill.

Drabble, J. 2000. *An Economic History of Malaysia, c.1800–1990: the Transition to Modern Economic Growth*. New York: Palgrave Macmillan.

Drew, E. and Foster, G. 1994. *Information Technology in Selected Countries: Reports from Ireland, Ethiopia, Nigeria, and Tanzania*. Tokyo: United Nations University.

DST (Department of Science and Technology) 2010. *Vision, Mission and Corporate Values*. Online and available at: http://www.dst.gov.za/about-us/vision-mission, accessed 12 March 2010.

Dyker, D. 1999. Foreign Direct Investment in Transition Countries: A Global Perspective. In Dyker, D. (ed.) *Foreign Direct Investment and Technology Transfer in the Former Soviet Union*. Northampton: Edward Elgar Publishing: 8–26.

Eaton, J. and Kortum, S. 1999. International Patenting and Technology Diffusion: Theory and Measurement. *International Economic Review, 40*: 537–570.

Estonica (Encyclopaedia about Estonia) 2009. *The Role of Banking in the Development of the Estonian Economy*. Online and available at: www.estonica.org, accessed 6 March 2010.

FAZ 2006. Empfehlungen des BDI für das China Geschäft. *Frankfurter Allgemeine Zeitung*. 19 May 2006: 13.

Feldman, M. and Kogler, D. 2008. The Contribution of Public Entities to Innovation and Technological Change. In Shane, S. (ed.) *Handbook of Technology and Innovation Management*. Chichester: Wiley: 431–460.

Ferreira, M., Tavares, A. and Hesterly, W. 2006. Evolution of Industry Clusters through Spin-offs and the Role of Flagship Firms. In Tavares, A. and

Teixeira, A. (eds) *Multinationals, Clusters and Innovation. Does Public Policy Matter?* Houndmills and New York: Palgrave: 87–106.

Fisher, R., Ury, W. and Patton, B. 2004. *Getting to Yes: Negotiating Agreement Without Giving in.* New York: Penguin.

Geroski, P. 1994. *Market Structure, Corporate Performance and Innovative Activity.* Oxford: Oxford University Press.

Giroud, A. 2006. Is Government Support Really Worth it? Developing Local Supply Linkages in Malaysia. In Tavares, A. and Teixeira, A. (eds) *Multinationals, Clusters and Innovation. Does Public Policy Matter?* Houndmills and New York: Palgrave: 179–198.

Granovetter, M. 1973. The Strength of Weak Ties. *American Journal of Sociology, 6:* 1360–1380.

Granovetter, M. 1982. The Strength of Weak Ties: A Network Theory Revisited. In Marsden, P. and Lin, N. (eds) *Social Structure and Network Analysis.* Beverly Hills: Sage: 105–130.

Grant, R. 1991. *Contemporary Strategy Analysis. Concepts, Techniques, Applications.* Cambridge: Basil Blackwell.

Grant, R. 1996. Prospering in Dynamically-Competitive Environments: Organizational Capability as Knowledge. *Organization Science, 7:* 375–387.

Hamar, J. and Stephan, J. 2006. Results of a Fieldwork Project. In Stephan, J. (ed.) *Technology Transfer via Foreign Direct Investment in Central and Eastern Europe. Theory. Method of Research and Empirical Evidence.* New York: Palgrave: 126–159.

Hansen, M. 1999. The Search-Transfer Problem: The Role of Weak Ties in Sharing Knowledge across Organization Subunits. *Administrative Science Quarterly, 44:* 82–111.

Hoekman, B., Maskus, K. and Saggi, K. 2004. *Transfer of Technology to Developing Countries: Unilateral and Multilateral Policy Options.* World Bank Policy Research Working Paper 3332. Washington: World Bank.

Hollander, S. 1965. The *Sources of Increased Efficiency: A Study of DuPont Rayon Plants.* Cambridge: MIT Press.

Holsapple, C. 2003. Knowledge and its Attributes. In Holsapple, C. (ed.) *Handbook on Knowledge Management.* Berlin: Springer: 165–188.

Hu, Z. and Khan, M. 1996. *Why is China Growing so Fast?* IMF Working Paper 96/75. Washington: International Monetary Fund

IDA (Industrial Development Agency) 2009. *ICT Ireland.* Online and available at: http://www.idaireland.com/ accessed 1 March 2010.

Ikiara, M. 2003. *Foreign Direct Investment (FDI), Technology Transfer, and Poverty Alleviation: Africa's Hopes and Dilemma.* ATPS Special Paper Series, 16. Nairobi: African Technology Policy Studies Network.

IMF (International Monetary Fund) 2009. *World Economic Outlook Database.* Washington: International Monetary Fund.

Ismail, M. 1999. Foreign Firms and National Technological Upgrading: The Electronics Industry in Malaysia. In Felker, J. and Rasiah, R. (eds) *Industrial Technology Development in Malaysia. Industry and Firm Studies.* London and New York: Routledge Chapman & Hall: 21–27.

Javorcik, B. 2004. Does Foreign Direct Investment Increase the Productivity of Domestic Firms? In Search of Spillovers through Backward Linkages. *The American Economic Review, 94:* 605–627.

Jindra, B. 2006. The Theoretical Framework: FDI and Technology Transfer. In Stephan, J. (ed.) *Technology Transfer via Foreign Direct Investment in Central and Eastern Europe. Theory, Method of Research and Empirical Evidence.* Houndmills: Palgrave: 6–29.

Jones, J. and Wren, C. 2006. *Foreign Direct Investment and the Regional Economy.* Aldershot: Ashgate.

Kaplan, R. and Norton, D. 2001. *The Strategy-Focused Organization. How Balanced Scorecard Companies Thrive in the New Business Environment.* Boston: Harvard Business School Pres.

Kim, L. 1997. *Imitation to Innovation: The Dynamics of Korea's Technological Learning.* Boston: Harvard Business School Press.

Kim, L. 1998. Crisis Construction and Organizational Learning: Capability Building in Catching-up at Hyundai Motor. *Organization Science, 9:* 506–521.

Knight, K. 1963. *A Study of Technological Innovation: The Evolution of Digital Computers.* Pittsburgh: Unpublished Ph.D. Dissertation.

Kotler, P. 2000. *Marketing Management. The Millennium Edition.* London: Prentice Hall International.

Kuznets, S. 1966. *Modern Economic Growth: Rate, Structure, and Speed.* New Haven: Yale University Press.

Laar, M. 2007. *The Estonian Economic Miracle.* The Heritage Foundation, August 7. Online and available at: http://www.heritage.org/research/worldwidefreedom/bg2060.cfm accessed 1 March 2010.

Lall, S. 2003. Foreign Direct Investment, Technology Development and Competitiveness. Issues and Evidence. In Lall, S. and Urata, S. (eds) *Competitiveness, FDI, and Technological Activity in East Asia.* Cheltenham: Edward Elgar Publishing: 12–56.

Lebas, M. and Euske, K. 2002. A Conceptual and Operational Delineation of Performance. In Neely, A. (ed.) *Business Performance Measurement. Theory and Practice.* Cambridge: Cambridge University Press: 65–79.

Leutert, H.-G. and Südhoff, R. 1999. Technological Capacity Building in the Malaysian Automotive Industry. In Jomo, K., Felker, G. and Rasiah, R. (eds)

Industrial Technology Development in Malaysia. Industry and Firm Studies. London and New York: Routledge Chapman & Hall: 247–273.

Luecke, R. 2003. *Negotiation. Harvard Business Essentials.* Boston: Harvard Business School Press.

MacNeil, K. 2001. *The Wine Bible.* New York: Workman Publishing.

Malairaja, C. and Zawdie, G. 2004. The 'Black Box' Syndrome in Technology Transfer and the Challenge of Innovation in Developing Countries. The Case of International Joint Ventures in Malaysia. *International Journal of Technology Management and Sustainable Development, 3:* 233–251.

Meadows, D.H., Meadows, D.L., Randers, J. and Behrens, W. 1972. *The Limits of Growth.* London: Pan Books.

Mellahi, K., Frynas, J. and Finlay, P. 2005. *Global Strategic Management.* Oxford: Oxford University Press.

Mitchell, J. 1969. The Concept and Use of Social Networks. In Mitchell, J. (ed.) *Social Networks in Urban Situations. Analysis of Personal Relationships in Central African Towns.* Manchester: The University Press: 1–50.

Moeini, E. and Zawdie, G. 1998. Import Substitution, Technological Learning and Innovation in Strategic Industries in Iran: A Survey of Evidence. *Science, Technology, and Development, 16:* 17–43.

Morais da Costa, T., Videira, A. and Veloso, F. 2006. Assessing the Value Creation and Backward Linkages in Foreign Direct Investment: A Combination of Macro and Micro Tools. In Tavares, A. and Teixeira, A. (eds) *Multinationals, Clusters, and Innovation. Does Public Policy Matter?* New York: Palgrave: 214–233.

Moran, T. 2005. How does FDI Affect Host Country Development? Using Industry Case Studies to make Reliable Generalization. In Moran, T., Graham, E. and Blomström, M. 2005: *Does Foreign Direct Investment Promote Development?* Washington: Institute for International Economics & Centre for Global Development: 281–313.

Moran, T., Graham, E. and Blomström, M. 2005. Conclusions and Implications for FDI Policy in Developing Countries, New Methods of Research, and Future Research Agenda. In Moran, T., Graham, E. and Blomström, M. 2005. *Does Foreign Direct Investment Promote Development?* Washington: Institute for International Economics & Centre for Global Development: 375–395.

Morgan, J. 2002. *Research and Experimental Development Statistics 2000.* Newport: Office of National Statistics.

Müller, T. and Schnitzer, M. 2003. *Technology Transfer and Spillovers in International Joint Ventures.* Munich Discussion Paper 2003–22. University of Munich.

Müller-Stewens, G. and Lechner, C. 2005. *Strategisches Management. Wie strategische Initiativen zum Wandel führen.* Stuttgart: Schäffer-Poeschel Verlag.

Mutch, A. 2008. *Managing Information and Knowledge in Organizations*. New York and London: Routledge.

Navaretti, G. and Bigano, A. 1998. R&D Inter-firm Agreements in Developing Countries. Where? Why? How? In Navaretti, G. and Bigano, A. (eds) *Creation and Transfer of Knowledge. Institutions and Incentives*. Berlin: Springer: 33–62.

Nonaka, I. 1994. A Dynamic Theory of Organizational Knowledge Creation. *Organization Science, 5:* 14–37.

O'Malley, E. and Hewitt-Dundas, N. and Roper, S. 2008. High Growth and Innovation with Low R&D: Ireland. In Edquist, C. and Hommen, L. (eds) *Small Country Innovation Systems: Globalization, Change and Policy in Asia and Europe*. Northampton: Edward Elgar Publishing: 156–193.

OECD (Organisation for Economic Co-operation and Development) 1998. *Policy Evaluation in Innovation and Technology: Towards Best Practices*. Paris: OECD.

OECD (Organisation for Economic Co-operation and Development) 2009. *OECD Factbook 2009. Economic, Environmental, and Social Statistics*. Paris: OECD.

Osterloh, M. and Frey, B. 2002. *Successful Management by Motivation. Balancing Intrinsic and Extrinsic Incentives*. New York: Springer.

Phillips, F. 2001. *Market-Oriented Technology Management. Innovating for Profit in Entrepreneurial Times*. Berlin: Springer.

Polanyi, M. 1966. *The Tacit Dimension*. New York: Peter Smith Publications.

Porter, M. 1985. *Competitive Advantage: Creating and Sustaining Superior Performance*. New York: The Free Press.

Porter, M. 1990. *The Competitive Advantage of Nations*. New York: The Free Press.

Rehäuser, J. and Krcmar, H. 1996. Wissensmanagement im Unternehmen. In Schreyögg and Conrad, P. (eds) *Managemenforschung 6. Wissensmanagement*. Berlin: Walter de Gruyter: 1–40.

Ricken, B. 2005. *Entwicklung eines Instrumentes zur Analyse und Steuerung informaler Organisationsstrukturen*. München and Mering: Rainer Hampp Verlag.

Ricken, B. and Seidl, D. 2010. *Unsichtbare Netzwerke: Wie sich die soziale Netzwerkanalyse für Unternehmen nutzen lässt*. Wiesbaden: Gabler Verlag.

Rivoli, P. and Salorio, E. 1996. Foreign Direct Investment and Investment under Uncertainty. *Journal of International Business Studies, 27:* 335–357.

Saggi, K. 2000. *Trade, Foreign Direct Investment, and International Technology Transfer: A Survey*. World Bank Policy Research Working Paper 2349. Washington, DC: World Bank.

Sanchez, R. 2001. Managing Knowledge into Competence: The Five Learning Cycles of the Competent Organization. In Sanchez, R. (ed.) *Knowledge*

Management and Organizational Competence. Oxford: Oxford University Press: 3–38.

Sands, A. 2005. The Irish Software Industry. In Arora, A. and Gambardella, A. (eds) *From Underdogs to Tigers. The Rise and Growth of the Software Industry in Brazil, China, India, and Israel*. Oxford: Oxford University Press: 41–71.

Sebenius, J. 2002. The Hidden Challenge of Cross-Border Negotiations. *Harvard Business Review, March*: 76–85.

Shane, S. 2009. *Technology Strategy for Managers and Entrepreneurs*. Upper Saddle River: Pearson Prentice Hall.

Singapore Department of Statistics 2009. *Yearbook of Statistics*. Singapore: Singapore Department of Statistics.

Smith, J. 2009. *Science and Technology for Development*. London & New York: Zed Books.

Spangle, M. and Isenhart, M. 2003. *Negotiation. Communication for Diverse Settings*. Thousand Oaks: Sage.

Stopford, J. and Wells, L. 1972. Managing the Multinational Enterprise. New York: Longman.

Taylor, D. 2010. *Supply Chains – A Manager's Guide – Chapter 1*. Online: Business Intelligence. Available at: http://www.businessintelligence.com/extract. asp?id=4 accessed 12 March 2010.

TDNE (The Daily News Egypt) 2009. *Egypt's Textile Industry needs Bottom-Up Approach, say Experts*. Online and available at: http://www.thedailynewsegypt. com/article.aspx?ArticleID=21270 accessed 10 March 2010.

UNCTAD (United Nations Conference on Trade and Development) 2001. *Transfer of Technology*. New York: United Nations Publications.

UNCTAD (United Nations Conference on Trade and Development) 2003a. *Transfer of Technology for Successful Integration into the Global Economy*. New York: United Nations Publications.

UNCTAD (United Nations Conference on Trade and Development) 2003b. *Creating an Attractive Environment for Foreign Direct Investments (FDI). Best Practices*. New York: United Nations Publications.

UNCTAD (United Nations Conference on Trade and Development) 2003c. *Africa's Technology Gap. Case Studies on Kenya, Ghana, Uganda and Tanzania*. New York: United Nations Publications.

UNCTAD (United Nations Conference on Trade and Development) 2004. *Facilitating Transfer of Technology to Developing Countries: A Survey of Home-Country Measures*. New York: United Nations Publications.

UNCTAD (United Nations Conference on Trade and Development) 2005a. *World Investment Report. Transnational Corporations and the Internalization of R&D*. New York: United Nations Publications.

UNCTAD (United Nations Conference on Trade and Development) 2005b. *A Case Study of the Electronics Industry in Thailand*. New York: United Nations Publications.

UNCTAD (United Nations Conference on Trade and Development) 2009. *World Investment Report 2009. Transnational Corporations, Agricultural Production and Development*. New York: United Nations Publications.

UNIDO (United Nations Industrial Development Organization) 1995. *Finding Technology. Manual on Technology Transfer Negotiation*. Vienna: United Nations Publications.

UNIDO (United Nations Industrial Development Organization) 1996. *Negotiating. Manual on Technology Transfer Negotiation*. Vienna: United Nations Publications.

UN ESCAP (United Nations Economic and Social Commission for Asia and the Pacific) 2002. *Development of the Automotive Sector in Selected Countries of the ESCAP Region*. New York: United Nations Publications.

Urata, S, and Kawai, H. 2000. Intrafirm Technology Transfer by Japanese Manufacturing Firms in Asia. In Ito, T. and Krueger, A. (eds) *The Role of Foreign Direct Investment in East Asian Economic Development*. Chicago: University of Chicago Press: 49–77.

Varum, C. 2006. International Buyer-Supplier Relationships, Transfer of Knowledge and Local Suppliers' Capability. In Tavares, A. and Teixeira, A. (eds) *Multinationals, Clusters and Innovation. Does Public Policy Matter?* New York: Palgrave: 234–252.

Vertzberger, Y. 2005. *Strategic Policy Reform: Policy Resilience, Learning and Change*. Paper to the 46th Annual Convention of the International Studies Association, Honolulu, 1–5 March.

WEF (World Economic Forum) 2009. *The Global Competitiveness Report 2009–2010*. Geneva: World Economic Forum.

Wing, R. 1998. *The Art of Strategy: A New Translation of Sun Tzu's 'The Art of War'*. New York: Main Street Books.

WSJ (The Heritage Foundation and *The Wall Street Journal*) 2009. *Index of Economic Freedom*. http://www.heritage.org/index/.

WWEA (World Wind Energy Association) 2009. *World Wind Energy Report 2008*. Bonn: World Wind Energy Association.

Index